Integrated Circuits and Systems

Volume 68

Series Editor

Anantha P. Chandrakasan, Massachusetts Institute of Technology, Cambridge, MA, USA

For further volumes:
http://www.springer.com/series/7236

Keren Bergman · Luca P. Carloni
Aleksandr Biberman · Johnnie Chan
Gilbert Hendry

Photonic Network-on-Chip Design

Springer

Keren Bergman
Department of Electrical Engineering
Columbia University
New York, NY
USA

Luca P. Carloni
Aleksandr Biberman
Johnnie Chan
Gilbert Hendry
Columbia University
New York, NY
USA

ISSN 1558-9412
ISBN 978-1-4419-9334-2 ISBN 978-1-4419-9335-9 (eBook)
DOI 10.1007/978-1-4419-9335-9
Springer New York Heidelberg Dordrecht London

Library of Congress Control Number: 2013942140

Printed on acid-free paper

Springer is part of Springer Science+Business Media (www.springer.com)

Preface

This book is a product of an emerging interdisciplinary field that is using photonic communications to address many of the challenges associated with scaling computing system performance. With the emergence of multicore architectures and the ever-growing quest for parallelism, system performance is increasingly bound by its communication rather than computation capabilities. Data movement in the optical domain offers many unique advantages and the possibility of new interconnection network architectures that can enable future computing performance gains. The underlying photonic technologies also present many unique design challenges. It is our goal in this book to bridge the gap among these interdisciplinary fields and create a common reference for evolving the design and development of chip-scale photonic networks.

Optical communication, which already has major roles in large-scale computing systems, will be completely transformed by the silicon photonic chip-scale integration. In turn, silicon photonics could potentially become the most important technology to sustain the continued performance scaling of integrated circuits over the next decades.

The distinctive properties of photonic interconnects are explained in the first half of the book. This part also contains a comprehensive overview of the rich family of photonic devices that have been developed over the past few years, thanks to some fundamental engineering breakthroughs. In the second half of the book, three main classes of photonic network architectures are presented to illustrate the variety of design solutions that are made possible by the combination of photonic and electronic devices.

The two halves of the book are connected by the presentation of a design and simulation environment as a common toolset that fosters research collaboration in this emerging interdisciplinary field. On one hand, it enables computer system engineers to explore the use of new devices to design photonic network architectures. On the other hand, it enables photonic researchers to understand the impact of different device designs on system performance.

In the presentation of the various topics, it was our intention to balance theoretical models and simulation domains with empirical results. This way the reader can develop a working knowledge of what is theoretically possible and what has already been successfully demonstrated. We hope that researchers, designers, and architects across disciplines find the material in this book relevant and gain an understanding and appreciation of this exciting field.

New York, May 2013

Keren Bergman
Luca P. Carloni
Johnnie Chan
Aleksandr Biberman
Gilbert Hendry

Contents

Chapter 1
Introduction

1.1 Transistors to Photonics

Over the past four decades the progress of computing systems was largely dominated by the underlying acceleration in microprocessor performance and extraordinary advances in semiconductor technology. Improved fabrication methods and increasing die sizes were manifested in Moore's law, predicting in 1965 that the number of transistors integrated on a single die will be roughly doubled every two years [1]. Along with additional advances in circuit design techniques and processor microarchitectures, these improvements led to rapidly increasing clock speeds and to the extremely-high performance presented by CMOS-based microprocessors.

The past trend of continuous acceleration in single microprocessor performance has undergone a major paradigm shift in recent years as limits on power dissipation have impeded the continued scaling in single processor speeds and led to the emergence of multicore architectures and chip multiprocessors (CMPs). Multicore architectures optimize performance-per-watt by operating multiple parallel processors at lower clock frequencies. Many commercial chips must restrict the number of processing cores that can operate simultaneously to avoid overheating. The amount of data that can be transferred between processor chips and memory chips is often limited by power cost. Power also limits the number of chips that can be hosted on a single embedded computing board and the amount of data that can be processed by the server of a cloud-computing cluster. Energy efficiency has clearly become a key metric in the design of future computing platforms.

Performance scalability of microprocessors and the multicore architectures of CMPs are becoming increasingly constrained by limitations in power dissipation, chip packaging, and the data throughput achievable by the on- and off-chip interconnection networks. To address the continued performance scalability of future CMPs, three critical interconnect-centric challenges clearly emerge:

- Global communication among the processing cores consumes an increasing portion of the limited on-chip power budget, thus impeding future performance gains.

K. Bergman et al., *Photonic Network-on-Chip Design*,
Integrated Circuits and Systems, DOI: 10.1007/978-1-4419-9335-9_1,

- The power dissipation problem is greatly exacerbated for off-chip electronic interconnects as they typically consume at least one order of magnitude more power even for short distances and do not scale significantly with new technology nodes.
- The off-chip communication bottleneck, a major challenge in current CMPs due to limited on-chip power budget and pinout, becomes a scaling barrier to memory bandwidth and system-wide data movement.

The interconnection networks of the CMPs have thus become a substantial determinant to overall system performance, since they serve as the communication links between pairs of cores, and provide the means to connect cores to off-chip inputs/outputs (I/O) and memory. Interconnection networks are being designed with larger datapath widths and higher signaling frequencies to meet the requirements of certain communication-bound applications. However, the power dissipation of electronic links tends to scale with throughput performance, causing a remarkable increase in overall chip power dissipation. These performance trends, combined with the thermal limitations of current chip-packaging technologies, have created the challenge of finding new technological solutions that can supply enough bandwidth to all the processing cores while maintaining a sustainable power dissipation level. One such technology is *integrated photonics* which has been slated with the potential to mitigate the many challenges associated with on- and off-chip electrical interconnection networks.

This book addresses the usage and integration of silicon nanophotonics for future computing systems.

Nanophotonics is the key to bringing integrated optical communications to computing systems. Silicon nanophotonics, enabled by their small device footprints, ultra-low capacitances, and the tight proximity of electronic drivers, offer the possibility of generating and receiving optical signals with fundamentally superior energy efficiencies. The insertion of photonic interconnection networks further changes the energy scaling rules: once a photonic path is established, the data is transmitted end-to-end without the need for power consuming repeaters, regenerators, or buffers.

The introduction of optical communications in computing systems at all scales will allow applications to leverage the advantages that have already revolutionized telecommunication systems: extremely high bandwidth density with minimal latencies, high energy efficiency per unit of bandwidth, and immunity to electro-magnetic effects such as noise and crosstalk. Bandwidth density measures the data throughput through an area or volume. Higher bandwidth density is particularly attractive from a deployment and integration standpoint since it enables greater throughput within equivalent or smaller physical dimensions. Integration of nanophotonics-enabled connectivity can occur at all levels: making the communication between rack servers independent of their relative position in distributed data centers, overcoming the communication bottleneck between processors and memory chips on a single board; and providing unique communication capabilities among the multiple processing cores of future chips.

The inter-core communications in a CMP is accomplished via an interconnect subsystem, referred to as a network-on-chip (NoC). Networks-on-chip have been introduced to reduce the wiring complexity, by designing regular topologies that can achieve predictable bandwidth, latency, and power dissipation from communication between the cores. These NoCs are analogous to modern telecommunications networks, having nodes interconnected by routers that direct packets of information from a source core to a destination core. Commercial implementation of NoCs appeared in the Tilera TILE microprocessor series [2] and the Cell Broadband Engine [3]. Thus far, electronic-enabled interconnects have been able to satisfy the communication requirements of current CMPs. However, as these systems continue to scale in performance and size, it becomes increasingly difficult to maintain a network that can both accommodate the communication demands and stay within power dissipation limits of the system package [4, 5]. Electronically-enabled interconnects in CMPs already account for over 50 % of the dynamic power dissipated in some high-performance chips [6]. The portion of dissipated power that comes from the interconnect is expected to continue to grow with time and will become the limiting factor in performance scaling again.

To maximize the available communication bandwidth in these networks-on-chip, interconnecting wires are typically connected in parallel, forming short communication links that can achieve higher total throughput. With increasing chip area and power dissipation constraints imposed by a growing number of cores, these communication buses are increasingly limited in how many wires they can feasibly sustain and in the speed of each wire, severely limiting the total utilizable bandwidth on chip. With limited bandwidth, designing the networks-on-chip requires a stringent balance between the available resources. Careful provisioning of communication access between the cores, on-chip cache memory, and off-chip memory interfaces must be considered in order to ensure maximum utilization and system performance.

Off-chip bandwidth between the cores and memory is limited, in what is commonly referred to as the *processor-memory performance gap*, which grows exponentially with every new processor generation. This disparity between the processor and memory has resulted from the annual performance improvement rate of 60 % for the processor, and access time reduction of less than 10 % for the memory [7]. The resulting interplay between off-chip bandwidth and access latency reduces the performance of the chip multiprocessor, asymmetrically affecting memory-intensive applications, and is the primary obstacle in achieving performance gains in computing systems.

For the performance scaling trends to continue, a paradigm shift must take place in the way that computer architectures are built and designed. The shift can either be brought about through fundamental changes to the way that computation logic is devised, or alternatively, and more dramatically, through a migration in underlying technology. One such potentially paradigm-shifting solution is the usage of optics, or more specifically *nanophotonic interconnect technologies*.

Fig. 1.1 Illustration of a current typical interconnect architecture which utilizes communication buses composed of parallel electronic wires

1.2 Photonics for Memory

The relatively-distance-independent and high-data-rate nature of optics is well suited for another major challenge facing computing systems: main memory interconnect architectures. The communication link between the CPU and main memory is a critical performance bottleneck for current fully electronic computing systems. Figure 1.1 illustrates the current typical structure of a memory subsystem. The component which we might typically think of as the *processor* is the integrated circuit (IC) which handles the pipeline that performs arithmetic functions, logic functions, issues requests for data retrieval from memory, and issues requests for data transmission to memory. The memory *controller* translates processor memory requests into logic signals necessary to access the requested memory elements. The *memory* itself in typical commercial systems is arranged as dual in-line memory modules (DIMMs), which are daughter cards mounted with several memory chips (typically dynamic random-access memory, also known as DRAM).

When a cache miss occurs in a traditional memory hierarchy, the processor must communicate with main memory to access the addressed data. This action requires a communication process to transpire off the chip, representing a domain boundary traversal and troublesome engineering challenge for system architects. Memory interactions occur as follows. First, the processor issues a request to the memory controller. Next, the memory controller must translate the request into the proper signaling to interact with the addressed memory cells. Lastly, the memory honors the commands from the memory controller and performs the requested action. In the event of a memory read operation, the data must be sent back through the controller and then to the processor. As is seen in the illustration (Fig. 1.1), the connection between the processor and the memory controller, and the memory controller and main memory requires a signaling bus that is composed of many wires in parallel. While many current commercial processors possess integrated memory controllers, this does not preclude the need for a wide bus to communicate with memory. Current third generation double data-rate (DDR3) DRAM requires 240 pins for proper electrical signaling. Memory systems have successfully been able to scale in capacity, but not without additional complexity costs from wider signalling buses and stricter timing requirements. This increased wiring complexity hinders improvements in bandwidth and latency.

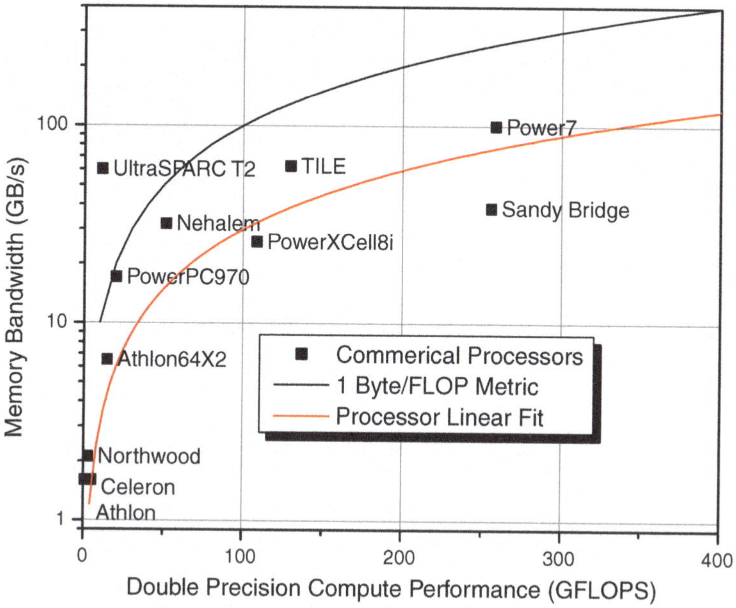

Fig. 1.2 Memory performance of commercial micro-processors in recent years. [Data was compiled from publicly available documents and publications. In instances where no memory pin count number was available, a third of total pin count was assumed]

Conventional computer architecture designs are able to alleviate issues of memory access latency by leveraging temporal and spatial locality of data. The presence of data locality enables the utilization of caching systems to hide access latencies. However, new cluster computing application classes have emerged in recent years which no longer conform to temporal or spatial locality assumptions. This results in much greater sensitivity to the throughput and latency of memory. A common metric that architects are increasingly specifying for a properly designed computer system is one byte (B) I/O transferred per floating-point operations (FLOP). In other words, 1 B/FLOP specifies a balance between memory bandwidth and computation performance. This metric is regarded as a rule-of-thumb for what is required of memory intensive applications. Figure 1.2 shows the recent trend in computational performance versus the available memory bandwidth. The plot shows a trend in commercial processors that is half an order of magnitude below the 1 B/FLOP metric. This requirement for constant streams of large amounts of data effectively nullifies the performance that caching can bring.

Current memory sub-systems place the DIMM components near the CPU. The reason for the close proximity is to reduce delay and increase frequency cutoff in the wire traces. However, a design tradeoff arises from the need to meet capacity demands by including many DIMMs which conflicts with the available area when the traces are limited in length. Optics can eliminate distance-dependent performance. By enabling

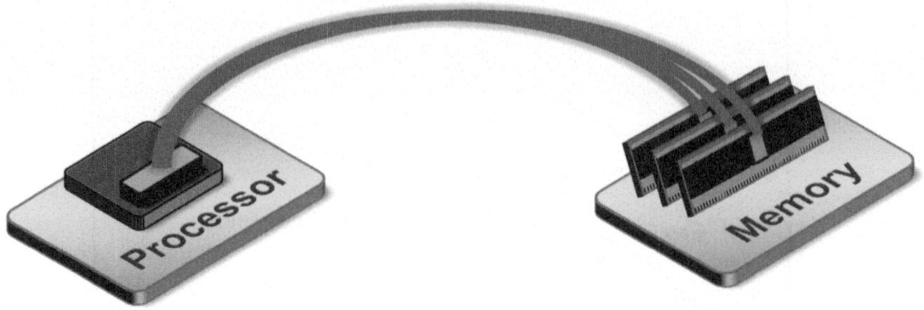

Fig. 1.3 Illustration of an optically-attached memory compute system with a processor attached to a single memory bank (composed of multiple DIMMs) via an optical bus

optical memory links, memory can be placed at farther distances while maintaining high data rates.

The problem of memory access is further exacerbated when considering multiple processors. At the rack and cluster scale, cache coherency is no longer feasible due to the extensive overhead incurred by the disparate locations of processing nodes. Multiprocessor systems at these scales take on a non-uniform memory access (NUMA) characteristics as computer architects try to constrain most communications to local memory. This constraint is predominately imposed due to the diminishing amount of available communication throughput as data moves farther and farther away from the processing core. This reduction in throughput is referred to as a bandwidth taper and can result in an order of magnitude difference between local and global memory access [8]. While NUMA architectures have been used to much success, new application classes that require large memory capacities (which will necessarily need to be physically located across the system) and external I/O will upend this assumption. By introducing optical technology, the restrictive memory bandwidth taper can potentially be eliminated.

The advantages that optics can leverage naturally make it an ideal technological solution to the challenges facing memory for computing. Research has shown that the enabling of optically-attached memory can provide significant performance advantages for typical high-performance computational algorithms. Figure 1.3 shows a hypothetical optical link between a processor and DIMMs. The processor and DIMMs each have integrated photonic transceiver components. This close integration of electronic logic and photonic components is key to eliminating the need for board-level wire traces and consequently the delay characteristics of off-chip communications. IBM Research has experimentally demonstrated this tight integration of photonics with electronic drivers [9].

A potential extension of the optically-attached memory is the optical-network attached memory which places an optical network between the processor and memory. This enables the possibility of utilizing multiple memory banks for each processor chip. Additionally, the versatility of an optical network can also be utilized to connect with other forms of I/O such as sensors, interfaces, and networks.

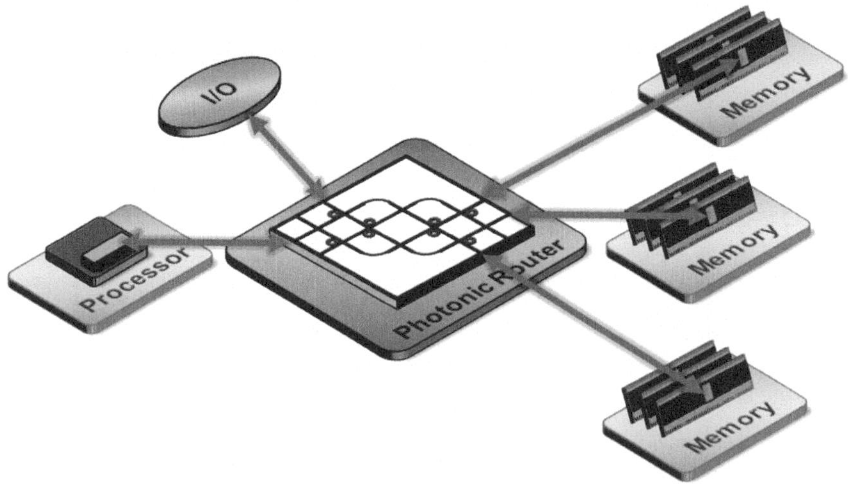

Fig. 1.4 Illustration of a compute system with a single processor attached to multiple memory banks and external I/O via a photonic interconnection network

The bandwidth density offered by optics enables the creation of such systems. Figure 1.4 visualizes this concept of attaching memory and I/O with an optical router.

One issue that the memory subsystems of current computer systems face is in the available off-chip I/O bandwidth. While on-chip bus bandwidths can reach terabits-per-second scales, off-chip memory bandwidths are orders of magnitude less at hundreds of gigabits-per-second. For example, the Tilera Tile processor is a 64-core chip arranged in an 8×8 mesh configuration with 2.56 Tb/s of bisection bandwidth and an off-chip memory bandwidth of 200 Gb/s [10]. This is primarily a limitation of the available pin count on chip packaging. Current state-of-the-art chips contain a maximum of around 2000 pins, with a significant number of the pins being utilized for power delivery and grounding.

Figure 1.5 plots estimates of the number of pins that are devoted to I/O for a sample set of processors (red squares) in the past decade. Figure 1.5 also shows the targeted number of pins in the next decade which are values published by the International Technology Roadmap for Semiconductors (ITRS) in 2010 [11]. Lastly, the figure also shows the required pin count for each processor if it were to achieve the 1 B/FLOPS metric, with estimated scaling of the clock frequency and improvements in processor performance. Notable is the fact that current commercial processors approximately follow the trend expected by the ITRS, however, this trend is almost an order of magnitude lower than the required pin count for 1 B/FLOPS performance. This electronic packaging problem is a potential area where photonics can bring about significant improvement to current architectures.

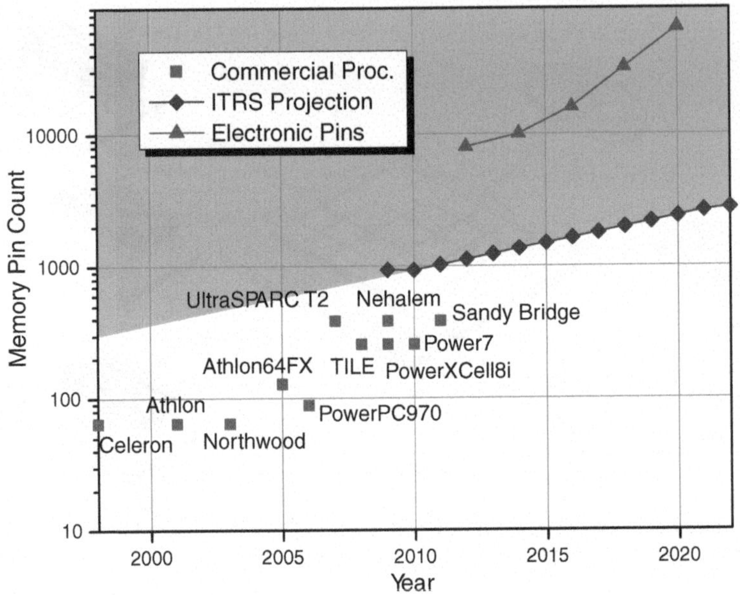

Fig. 1.5 Processor I/O pin scaling of commercial micro-processors in recent years (estimated, *red square* markers). Plot also shows ITRS projections for targeted pin count in the next decade (*blue diamond* markers), and the required pin count of a processor package in order to achieve a performance of 1 B/FLOP (*green triangle* markers) [Data was compiled from publically available documents and publications. In instances where no memory pin count number was available, a third of total pin count was assumed]

1.3 Remainder of this Book

Photonic interconnection networks offer solutions to many of the challenges associated with scaling the performance of the computing system from single-chip multiprocessors, to board-scale processor-memory systems. However, many solutions still need to be fathomed at both the device level and system/architecture level. This book describes all the technologies that have thus far been developed towards this effort and discusses some of those that still need to be developed. The contents of this book can be thought of as a comprehensive blueprint towards the realization of photonic interconnection networks.

The engineering challenges this book sets out to address arise in three technological domains: (1) devices, (2) tools, and (3) architectures. Within the device realm, physicists must design, create, and utilize novel components for enabling the fundamental functions of an optical link. On the opposite side of the spectrum are the computer architects, who must create systems from the combination of fundamental devices. Lastly, the domain that welds these two opposite but closely intertwined fields together are the tools, which must be designed and created in order to facilitate the collaborative and cohesive progress of the two areas.

Chapter 2 introduces the framework for a canonical photonic communication link. This spans the logical blocks required for optical message generation, transportation, and reception.

Chapter 3 describes all the basic devices necessary for the creation of each segment of the photonic link and it overviews the fabrication technology required to produce the devices. One device that is emphasized is the microring resonator which has extremely versatile usage properties. Other alternative components are also described.

A methodology for design and analysis of photonic network architectures is presented in Chap. 4.

Chapters 5–7 describe three different classes of photonic network architectures together with their advantages and disadvantages, illustrated through case studies.

Finally, concluding remarks are presented in Chap. 8.

References

1. G. E. Moore, "Cramming more components onto integrated circuits", *Electronics,* vol. 38, no. 8, pp. 114–117, Apr. 1965.
2. S. Bell, B. Edwards, J. Amann, R. Conlin, K. Joyce, V. Leung, J. MacKay, M. Reif, L. Bao, J. Brown, M. Mattina, C.-C. Miao, C. Ramey, D. Wentzlaff, W. Anderson, E. Berger, N. Fairbanks, D. Khan, F. Montenegro, J. Stickney, and J. Zook, "TILE64 processor: A 64-core SoC with mesh interconnect", in *Solid-State Circuits Conference (ISSCC), 2008. IEEE International,* Feb. 2008, pp. 88–598.
3. S. Clark, K. Haselhorst, K. Imming, J. Irish, D. Krolak,, and T. Ozguner, "Cell broadband engine interconnect and memory interface", in *Hot Chips 17,* Aug. 2005.
4. J. Meindl, "Interconnect opportunities for gigascale integration", *Micro, IEEE,* vol. 23, no. 3, pp. 28–35, May-Jun. 2003.
5. R. Ho, K. Mai, and M. Horowitz, "The future of wires", *Proceedings of the IEEE,* vol. 89, no. 4, pp. 490–504, Apr. 2001.
6. N. Magen, A. Kolodny, U. Weiser, and N. Shamir, "Interconnect-power dissipation in a microprocessor", in *Proceedings of the 2004 international workshop on System level interconnect prediction (SLIP),* Feb. 2004, pp. 7–13.
7. D. Patterson, T. Anderson, N. Cardwell, R. Fromm, K. Keeton, C. Kozyrakis, R. Thomas, and K. Yelick, "A case for intelligent RAM", *Micro, IEEE,* vol. 17, no. 2, pp. 34–44, Mar.-Apr. 1997.
8. S. L. Graham, M. Snir, and C. A. Patterson, *Getting Up to Speed, The Future of Supercomputing.* The National Academies Press, 2006.
9. J. Rosenberg, W. M. Green, A. Rylyakov, C. Schow, S. Assefa, B. G. Lee, C. Jahnes, Y. Vlasov, "Ultra-low-voltage micro-ring modulator integrated with a CMOS feed-forward equalization driver", in *Optical Fiber Communication Conference.* Optical Society of America, Mar. 2011, p. OWQ4.
10. D. Wentzlaff, P. Griffin, H. Hoffmann, L. Bao, B. Edwards, C. Ramey, M. Mattina, C.-C. Miao, J. F. Brown III, and A. Agarwal, "On-chip interconnection architecture of the Tile processor", *IEEE Micro,* vol. 27, no. 5, pp. 15–31, Sep. 2007.
11. International technology roadmap for semiconductors: 2010 report. [Online]. Available: http://www.itrs.net

Chapter 2
Photonic Interconnects

This chapter describes the most important characteristics and performance metrics of chip-scale communications. Figure 2.1 illustrates the general structure of all optical communication channels, which comprises of the communicating nodes and the optical link itself. The optical link consists of three functional elements: (1) generation, (2) routing, and (3) reception. *Generation* happens near a source node and involves the creation of a waveform in the optical domain for transporting useful information. *Routing* is for controlling the movement of optical data so that the useful information can travel from the source node to the destination node. Lastly, *reception* enables the optical link to translate the useful information back into the electrical domain to be used by the computing resource at the destination node. These three components (generation, routing, and reception) encompass everything needed for optical communications.

Although the high-level functional diagram of the typical optical link (Fig. 2.1) is similar to electronic interconnects, the two technology domains actually require fundamentally different design methods. In terms of generation and reception, the link requires a translation of electrons in the electrical domain to and from photons in the optical domain. Additionally, optical links can uniquely leverage wavelength-division multiplexing (WDM), which is the ability to simultaneously transmit multiple streams of data on a single physical communication medium by leveraging several optical carriers.

2.1 Photonic Technology

An introductory discussion of the underlying photonic technology is presented here. Further details of silicon photonics are presented in Chap. 3.

While there is a large variety of photonic devices being developed and researched, the actual number of types of devices required for an optical network is fairly small. The devices required for a photonic interconnection network include waveguides, couplers, modulators, detectors, switches, and filters. In this section, silicon photonic

K. Bergman et al., *Photonic Network-on-Chip Design*,
Integrated Circuits and Systems, DOI: 10.1007/978-1-4419-9335-9_2,

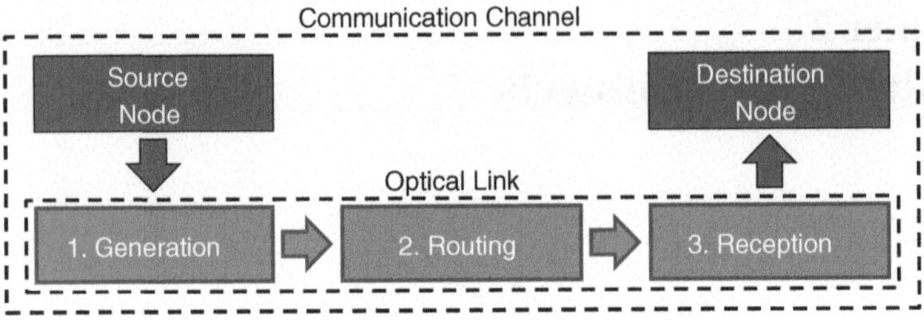

Fig. 2.1 High-level functional block diagram of all optical communication channels

devices will be described, elucidated in terms of their usefulness towards chip-scale optical networks, and contrasted with electronic equivalents.

2.1.1 Wavelength-Division Multiplexing

Parallel data streams in an electronic network require multiple spatially parallel wires. In contrast, a single waveguide (the photonic equivalent of a wire) can transport several streams of optical data by utilizing a unique wavelength for each independent data stream. In most scenarios, each optical carrier with its unique wavelength will not interfere with any other signal flowing along the same waveguide. WDM can be utilized to dramatically increase communication throughput and communication distance. WDM is a fundamental reason for why optical communications are an attractive solution for future high-performance systems. The utilization of WDM in photonic architectures will be discussed in later chapters.

The extent to which an optical device can leverage WDM is dictated by its spectral bandwidth property. Spectral bandwidth differs from network bandwidth. Spectral bandwidth is a measure of the range of frequencies, while network bandwidth (also known as throughput) measures the amount of digital information being transferred. Although the two metrics measure different properties, they are related. With more spectral bandwidth comes the ability to transmit more data.

Optical components are typically characterized by extremely large spectral bandwidths. For instance, a silicon waveguide can have a spectral bandwidth of hundreds of nanometers. This means that the waveguide supports the transmission of a digital signal at any center frequency within its spectral bandwidth range with a reasonable amount of loss. However, a single 10-Gb/s data signal exhibits a spectral bandwidth of less than one nanometer. While higher rate serial data signals can also be utilized (e.g. 40 Gb/s and 100 Gb/s), they are also unable to exhaust the hundreds of nanometers that are available.

The trick to effectively utilize the available spectrum is to *spectrally multiplex* many signals together. Multiplexing is the act of combining multiple independent signals together into a *WDM signal*. Demultiplexing is the act of separating out

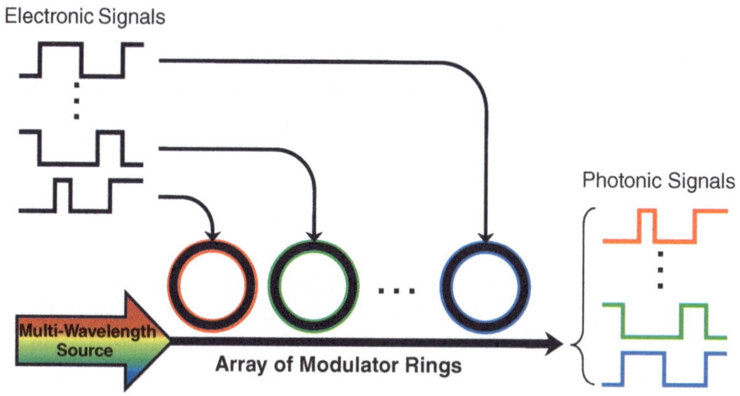

Fig. 2.2 Schematic of the conversion process between the spatially parallel electronic domain and wavelength-parallel optical domain

each serial data signal that comprises a WDM signal. Each signal would need to occupy mutually exclusive regions of the spectrum to avoid interference, but would be able to travel through the optical link simultaneously and in a single medium. This technique of aggregating multiple data signals together is known as wavelength division multiplexing.

One method to generate a WDM signal is to place multiple optical modulators in series along a single waveguide (Fig. 2.2). Each modulator is tuned to a different wavelength of light, thereby preventing any particular ring from interfering with the optical data generation of another ring. This scheme is discussed in detail in Sect. 3.4.2.

2.1.2 *Waveguides*

Waveguides can be regarded as the photonic equivalent of a wire. Waveguides are passive components which provide the physical links between all sources and destinations and enable connectivity between all photonic devices. Although they are simple devices, they are fundamental elements in all parts of an optical link. A photonic signal experiences attenuation (i.e. power loss) as it propagates through the waveguide. Insertion loss is the amount of attenuation a device induces if it is *inserted* into an optical link. For a waveguide, insertion loss is caused by free carrier absorption, light scattering at sidewall imperfections, and substrate leakage [1]. Proper routing of data requires waveguide portions to be straight, as well as bend, and to cross. Each bend and crossing of the waveguide presents additional sources of loss which must be considered when determining overall scalability of a photonic network.

A straight waveguide segment is simplest in design and typically has the lowest loss when compared to any other waveguide variations (i.e. bends and crossings). A 2005 paper [1] surveyed waveguides published by research groups around the world, showing SOI-based waveguides with losses ranging from 2.4 up to 110 dB/cm. More recently, silicon waveguides with cross sectional areas of approximately 500 nm × 250 nm have demonstrated improved losses of 1–2 dB/cm [2, 3]. Lower losses can be achieved using more exotic waveguide structures such as etchless silicon waveguides that have been shown to have losses of 0.3 dB/cm [4]. In terms of other current CMOS compatible materials besides crystalline silicon, silicon nitride is also a possible option due to its extremely low loss characteristics (losses of 0.1 dB/cm, [5]).

Just like an electronic wire, waveguides need to trace out paths with both straight sections and bending sections in order to divert signals correctly. Bends are necessary for the proper routing of optical paths, but also introduce an additional source of attenuation. This excess attenuation is inversely related to the bending radius. Thus, a smaller bending radius produces a larger excess loss factor. The amount of loss has been experimentally measured to be 0.005 dB per 90° with bending radius of 6.5 μm [2]. In general, bending radii longer than 5 μm produce negligible excess loss in comparison to the propagation loss of the waveguide itself.

Waveguide crossings are inherently required in silicon-based on-chip topologies due to the 2-D planar nature of the technology platform. Crossings occur whenever two waveguides intersect and can exhibit both insertion loss and crosstalk which can have an impact on system scalability and performance. This is especially true of photonic networks constructed on the single-crystalline silicon material (Sect. 3.1), which includes the vast majority of proposed chip-scale photonic architectures. This is in distinct contrast with electronic interconnects. Arbitrarily allowing the crossings of two wires on a single plane would cause a short circuit. Since many topologies inevitably require a large number of waveguide crossings, it is important for these devices to exhibit both low insertion loss and low crosstalk. A 6 μm × 6 μm double-etched crossing design has been fabricated and tested, and was shown to have fairly low insertion loss at 0.16 dB and high crosstalk suppression at about −40 dB [6].

2.1.3 Ring Resonators

The ring resonator is an instrumental device in the construction of photonic interconnection networks due to its versatility in implementing a variety of networking functions, compact footprint, and CMOS compatibility [7–13]. Ring resonators can be utilized to create modulators, filters, and switches.

Ring resonators are waveguides that form a closed loop which can be designed to manipulate the flow of light in a way that enables network functionality. Light interacts with the rings at specific periodically-spaced wavelengths in the optical spectrum, called *resonant modes*. When a waveguide is properly positioned next to a ring resonator, lightwaves injected into the waveguide that are rejected by the ring (termed *off resonance*) are transmitted past the ring (Fig. 2.3a). Lightwaves that

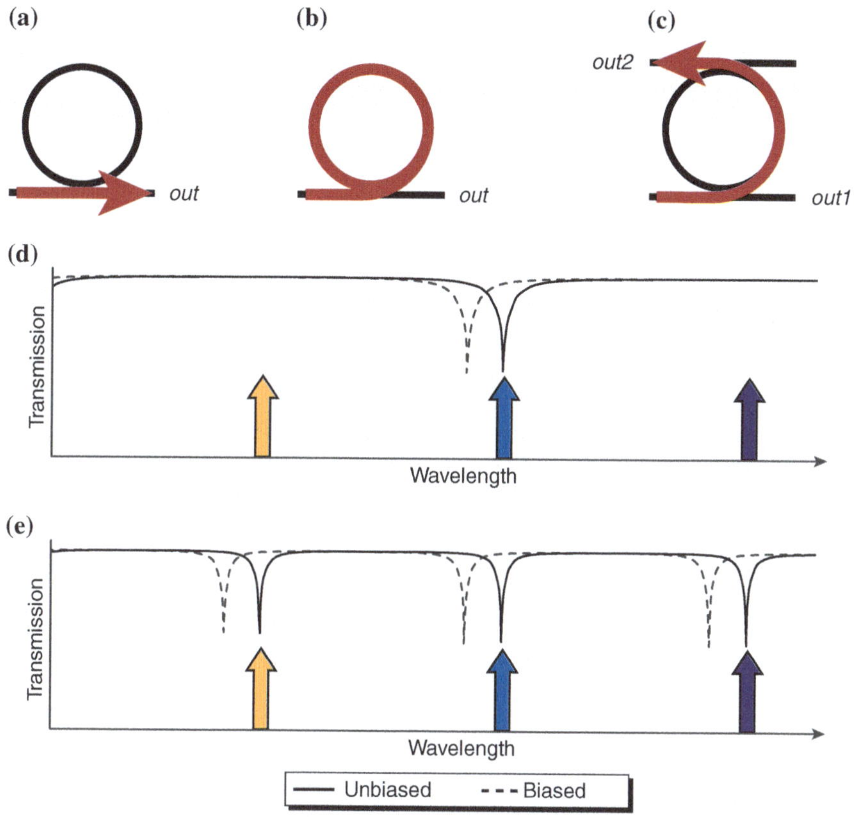

Fig. 2.3 Ring resonator functional characteristics. **a** Off-resonance wavelength with a single waveguide. **b** On-resonance wavelength with a single waveguide. **c** On-resonance wavelength with secondary waveguide. **d** Transmission spectra of a long FSR ring resonator. **e** Transmission spectra of a short FSR ring resonator. The *solid* and *dotted* spectra in **d** and **e** show the influence of electro-optic control on the resonances of the ring while in an electrically unbiased and biased state [14]

couple into the ring (termed *on resonance*) are not transmitted and are dissipated by the ring (Fig. 2.3b). Electro-optic rings can also be fabricated by properly doping the silicon material. This enhanced version of the ring can be electrically manipulated to fluctuate between the two resonance states to produce modulated light on the waveguide output. This modulated light provides the necessary mechanism to transform an electrical signal into an optical signal for the generation block of the optical link.

Alternatively, ring resonators can be designed to deliver on-resonance lightwaves onto a nearby secondary waveguide to enable filtering or switching functionality (Fig. 2.3c). Switching is a critical aspect of the transport and manipulation stage of the optical link as it allows a system to divert and control the path the optical signal takes in the network. Filters can be utilized in both the generation and reception stages as part of the multiplexing and demultiplexing of WDM signals. Filters can also be utilized in the transport/manipulation segment.

The *free spectral range* (FSR) of the ring resonator is inversely proportional to the circumference of the loop, and quantifies the spectral distance between wavelengths that couple and resonate with the ring. Modulators and filters which operate on a single wavelength ideally have a small circumference and large FSR, thereby allowing only a single on-resonance wavelength and rejecting all other channels (Fig. 2.3d). When filtering or switching is required on more than just one wavelength, a smaller FSR is desirable, so that several wavelength channels can be concurrently on-resonance with the ring (Fig. 2.3e). In this manner, the single ring resonator can be used to simultaneously manipulate all channels in a WDM signal with no additional cost in complexity or footprint.

Moreover, Fig. 2.3d, e illustrates how electro-optic control through free carrier injection can be used to manipulate the resonant wavelengths of the ring for modulation or active switching [9, 10]. Electrical manipulation can be accomplished by creating a P-I-N structure on the ring with the waveguide acting as the intrinsic region. Electrically biasing the P-I-N structure causes a shift in refractive index due to the free-carrier plasma dispersion effect in silicon [15]. This contrasts with thermal manipulation, which uses the thermo-optic properties of the material for index changes [16]. The diverse range in functionality and the controllability offered by the ring resonator make it an instrumental component in the design of photonic interconnection networks.

The FSR imposes a limitation on the number of wavelength channels that can be utilized in a WDM system. Ring resonator modulators should affect only a single wavelength channel, therefore the periodic nature of the resonances imposes an inherent limitation on the number of channels possible. Preston, et al. showed that a WDM interconnect based on ring resonators can maintain a satisfactorily low crosstalk level by having maximum wavelength channel count limitation of 62 when assuming 10-Gb/s datarates [17]. One cause of this limitation is that the minimum ring radius which can be fabricated also results in a maximum FSR limit of 50 nm. This issue can be addressed by exploiting more exotic resonator designs which can significantly elongate the FSR such as interferometric combining [18], photonic bandgap structures [19], and the Vernier effect [20]. These techniques can be used to increase the FSR, and correspondingly increase the available spectrum and allowable number of channels.

2.2 Photonic Metrics

The following three main performance metrics are utilized in the rest of this book.

Bit Error Rate (BER) is the fraction of bits detected by a receiver that do not match what the transmitter actually sent. For instance, due to presence of noise in a channel a receiver may detect a ‘1’ bit when the originating transmitter intended to send a ‘0’ bit. This metric can either be directly measured or analytically calculated. A value equal or close to zero (e.g. $<10^{-12}$) is said to be *error free*. BER is unitless.

Power Penalty (PP) is the reduction in signal-to-noise ratio (SNR) of a signal due to a change in the communication system. A PP may be caused by a variety of issues such as filtering and crosstalk. For example, replacing a high-quality modulator with a noisy modulator will add noise to the link, thereby decreasing the SNR and inducing a PP. The PP introduced into the link can be counteracted by increasing the signal power received by the receiver by an equal amount. This assumes that the noise power has no dependence on the signal power. Two ways to increase signal power at the receiver are to either reduce the insertion loss of the optical link or to increase the injected power of the laser. Signal and noise have units of watts, therefore SNR and PP are unitless values.

An **Eye Diagram**, which is directly observed on an oscilloscope, can be both a qualitative and quantitative tool to evaluate the temporal response of a digital data signal. After the digital data signal is received and sent to the oscilloscope, it is repetitively sampled to overlay and display a synchronized superposition of bit states and bit transitions on a single plot. The resulting pattern resembles a series of eyes between a pair of rails. From this pattern, it is possible to evaluate combined effects of channel noise, intersymbol interference, pattern dependencies, and even nonlinear optical effects in silicon. An open eye diagram corresponds to minimal signal distortion, and different noise effects cause the eye diagram to close.

2.3 Generation

The generation stage of an optical link converts digital information from the electronic domain into optical domain. From an ideal perspective, the only critical component to optical data generation is the modulator. However, from a practical standpoint the electronic data needs to be encoded and conditioned to be compatible with the modulation device and the rest of the optical network. Aside from compatibility, the generation stage also converts the stream of data into a format that can better leverage the advantages of optics.

Figure 2.4 illustrates the general structure of an optical generation stage. The first step is to encode incoming electronic data for error correction and signal conditioning purposes. The next step is to raise the transmission data rate as necessary through serialization. Typical electronic buses utilize multiple wires in parallel for transmission, therefore the serialization step can also lower the number of wires required at the output. Third, a driver circuit is required for each transmission wire to further condition each signal with the appropriate peak-to-peak voltage levels and to supply an adequate amount of current to drive the optical modulator.

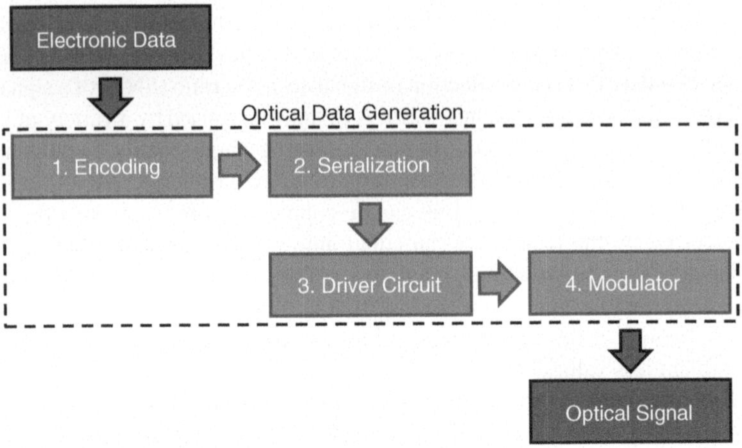

Fig. 2.4 Functional diagram of the generation stage of optical networks

2.3.1 Encoding

The encoding of a source stream of data should be purposefully done to generate a data stream that is compatible with critical design characteristics of the network and that is robust enough to recover from bit errors.

Transmission of data across a long distance requires *clock and data recover* (CDR) mechanisms at the reception component of the destination node. CDR can be accomplished when utilizing Manchester coding or some specially designed form of 8 b/10 b encoding scheme [21] to create enough transitions in the transmitted data so that a phase-locked loop (PLL) can recover the frequency and phase of the incoming data. Without a specialized coding schemes, the possibility of a long string of '0's or '1's arises which can allow the PLL to drift in phase, causing the reception component to lose the ability to recover the data.

In addition to providing CDR, the aforementioned coding schemes also provide DC balance for the signal. The ring-based modulators described in Sect. 3.4 are susceptible to performance degradation from temperature change. This is due to the spectral resonance having a strong dependence on index of refraction, which is a function of absolute temperature. One source of thermal perturbation is the signal driving the electro-optic ring modulator. If the DC value of the input signal constantly fluctuates in time, then the current through the P-I-N junction of the ring modulator also changes. This time-dependent change in current translates to a constantly changing thermal output which may destabilize the operation of the ring.

Lastly, error detection and correction methods should be implemented to satisfy quality-of-service requirements of the interconnect system. This is a necessary requirement for not only photonic networks, but all networks in general. Forward-error correction (FEC) is a technique where redundant bits are added to the data stream to add resilience to noise. A simple example of a FEC code is the Hamming code, which is a binary linear code that adds three additional check bits for every four data bits.

2.3.2 Serialization

The performance of real networks are governed by the deliverable latency and data rate. A figure of merit for characterizing the quality of a link is its *bandwidth-distance product*, which is defined as the product of the length of a link and the maximum deliverable datarate at that length. This figure is approximately constant as you change distance and maximum datarate. In other words, there is an inversely proportional relationship between the length of the link and the maximum datarate that can be transferred.

The *insertion loss* is a measure of the attenuation a signal receives from traversing a particular optical component. A device which has a positive insertion loss will output less power than it receives at its input. The spectral bandwidth refers to the acceptable range in frequencies of light that an optical component can effectively utilize. The dominant attribute of optics that has allowed it to proliferate in the telecom domain is its extremely high distance-bandwidth product due to its low insertion loss and high spectral bandwidth [22].

In contrast, high frequency electrical signals rapidly attenuate and are extremely inefficient at being transmitted over long distances. These properties make optics an attractive medium for transmitting high serialized datarates, which produce extremely high-datarate densities. In order for the network-on-chip domain to also leverage these optical advantages, systems should be set in place to increase the datarate of the serialized data.

In recent years, computation nodes have settled in terms of achievable clock frequency to somewhere between 2 and 4 GHz due mainly to power-dissipation constraints [23]. Optics, however, has been characterized to perform ideally at a much higher clock speeds, closer to 20–40 GHz. This clock mismatch can be accommodated by introducing a serialization circuit.

The task of a serializer is to up-convert an input data bus to a higher serialized datarate by combining multiple incoming data streams (i.e. wires). The aggregate datarate remains constant before and after the serializer. For instance, 128 wires each transmitting at 2 Gb/s can be serialized to 16 data streams each operating at 16 Gb/s. This tradeoff of number of wires versus transmission datarate has important implications when considering power dissipation requirements and available optical spectrum.

Serialization has reached widespread utilization in the electronics domain since it can physically reduce the size and number of wire channels. Commercial implementations of serialized data channels for system-to-system communications include RS-232, Universal Serial Bus (USB) [24], Serial ATA [25], and IEEE 1394 (Firewire) [26]. Examples of serializer circuits for on-chip and optical communications communications have also been developed and demonstrated [27, 28].

2.3.3 Driver Circuit

Drivers are necessary circuits for controlling the optical modulator device. This is due to differences between the electrical requirements of transistors used in digital logic and the electrical requirements of optical modulators. In an effort to reduce leakage current, current microprocessor manufacturers are moving towards lower operating threshold voltages for transistor technology. This trend runs counter to a P-I-N based optical ring modulator which operates based on a current swing instead of a voltage swing. A pre-amplifier circuit can be used to convert the incoming electrical signal to the higher voltage necessary to electrically actuate the device. Similarly, the transistor circuit may also not be suitable to deliver enough current, which can be remedied by an appropriately designed driver. Work at IBM has demonstrated such a CMOS driver circuit design for photonic modulators [29].

2.3.4 Modulator

The modulator translates the electrical signal into an optical signal so that it can propagate on the photonic interconnection network. Optical modulation requires three components: a light source, a medium for transporting the light, and a modulation device.

The light source is typically a laser which can be integrated into the same chip package as the modulator, or can be placed externally as a separate chip. An integrated laser source is referred to as an on-chip light source, while an external laser is referred to as an off-chip light source.

The modulated signal is generated on a waveguide and directed towards its destination. This book focuses exclusively on CMOS-compatible devices, therefore the integrated waveguides (ones that exist on-chip alongside the explored devices) considered will be composed of either silicon (Sect. 3.1.1) or silicon-nitride (Sect. 3.3.3).

Modulation devices come in a variety of flavors, varying in both physical mechanisms for modulation as well as material systems. This book focuses on the silicon material system since it is compatible with traditional CMOS foundries, and on ring-resonator-based modulators for their compact size, high-performance characteristics, and power efficiency. As described in Sect. 2.1.3, ring resonators can be designed to manipulate the flow of light in a way enables modulation.

2.4 Routing

The routing of data is essential for effectively moving data from the source node to destination node. A full treatment of the various forms of on-chip optical routing is described in Chaps. 5–7.

2.4.1 Optical Link

The simplest photonic interconnect is the *optical link* which utilizes no routing mechanism and only provides an all-optical connection between two nodes. Extensions of this type of optical interconnect beyond two nodes involve the usage of electronic repeaters and routers. This basic form of optical interconnect is comparable to active cabling used in current high-performance computing environments.

In what can be considered as the first step towards a full-scale photonic platform, Ophir et al. demonstrated a silicon-integrated optical link operating at a data rate of 3 GHz [30].

2.4.2 Wavelength Routing

Wavelength routing is used by network architectures that are constructed using ring-resonator-based filters which accordingly route lightwaves based on their wavelengths [31–34]. Any source node can address its intended destination through the selection of an appropriate transmission wavelength (i.e. source routing), which is then guided by the ring filters throughout the network. Transmission latencies can be designed to be extremely short when using wavelength routing since the propagation delay is simply the time of flight at the speed of light. However, spectral bandwidth is used for routing purposes which could have otherwise be used to increase communication data rates.

2.4.3 Spatial Routing

Network architectures that incorporate spatial routing use electro-optic broadband ring resonators to guide a large set of parallel wavelength channels along an optical path [35–37]. The ring resonators act as switches to simultaneously control the path of all incident wavelength channels. Spatial routing requires a priori establishment of the entire optical path which is typically created using a circuit-switching mechanism. While spatial routing exhibits longer latencies than wavelength routing due to the overhead of the circuit-switching protocol, it is able to leverage the entirety of the available optical spectrum for data striping to create extremely high bandwidth links. A previous study showed that the circuit-switching overhead can be amortized over large data messages, which is a characteristic of certain scientific applications typically executed on high-performance systems [36].

2.4.4 TDM Routing

Hendry et al. proposed the use of time-division multiplexing (TDM) to eliminate the circuit switching overhead associated with spatial routing [38]. *TDM routing* temporally divides the transmission medium into a continuous series of frames. Each frame is subdivided into several time slots which represent a different configuration of the entire optical network, and the set of all unique time slots completely connects all nodes in the network. The network is constructed using broadband ring switches, identical to the switches used for spatial routing, which are electro-optically reconfigured at the beginning of each time slot. A queued message at a source node will wait until an appropriate time slot arrives before it begins transmission, which contrasts with the spatial routing mechanism of immediately requesting the circuit allocation.

2.4.5 Wavelength-Selective Spatial Routing

Chan et al. proposed another novel on-chip photonic interconnect architecture that leverages a combination of wavelength-selectivity and spatial routing to increase path diversity within previously proposed spatial-routed photonic networks [39]. Spatial routing photonic network designs require circuit-switching style arbitration and can suffer from longer latencies and degraded bandwidth performance. These problematic issues are due to low path diversity and high contention probability caused by a fundamental architectural constraint that limits each physical optical path to a single communication link at any one point in time.

Traditional networks can leverage electronic virtual channels to statistically multiplex several logical links through a single physical electronic bus. However, the standard virtual channel technique requires buffering and processing which are not economically feasible in the photonic domain. Instead, wavelength selective-spatial routing (WSSR) utilizes spectral multiplexing to create several concurrent communication links with a single waveguide. Compared to traditional circuit-switched photonic networks, WSSR-enabled networks can achieve higher throughput performance with a minimal increase in design complexity and latency.

2.5 Reception

The final functional segment of an optical link is responsible for reverting the optical signal back into the electrical domain (Fig. 2.5). The principle behind optical reception can be described as the inversion of the processes that occurred in the generation of the optical signal. The optical signal is first received by a photo-detector, and then conditioned and decoded so that it can be processed by the destination node.

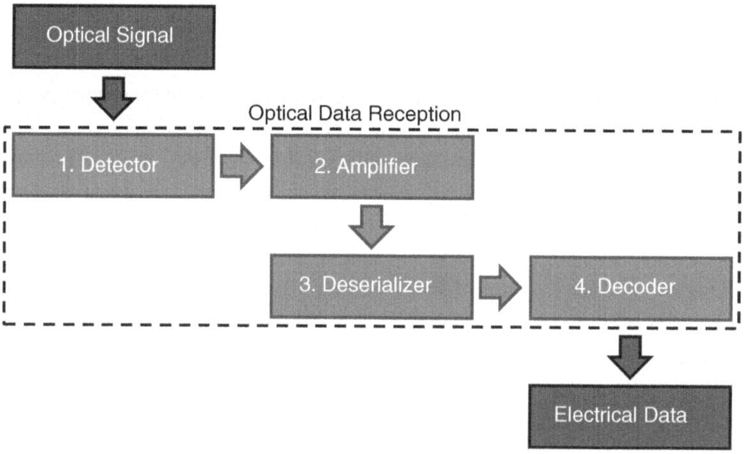

Fig. 2.5 Functional diagram of the reception stage of optical networks

2.5.1 Wavelength-Division Demultiplexing

Since WDM is extensively utilized to take advantage of the bandwidth density properties of optical communications, it is important to consider the technical requirements to properly receive a WDM signal. Typical detectors are sensitive to wavelengths of light across a large spectrum. Therefore, if a WDM signal were to be incident on the detector device, a superposition of all the channels would be received. Filters need to be placed before a detector in order to discriminate each channel of the WDM signal.

A complete receiver for a WDM signal with x number of channels requires x filters and x detectors. While the detectors can be identical because they are sensitive across a wide spectrum, each filter needs to be customized so that it is tuned to a unique channel.

2.5.2 Detector

The detector is responsible for converting the stream of photons into an electrical current. Detectors typically are constructed from a type of material that has an atomic characteristic which generates an electron when exposed to a photon. Detectors vary in performance with respect to efficiency, speed, and response spectrum. The reader is referred to Sect. 3.6 for detailed operational descriptions of the detector devices and its performance characteristics.

2.5.3 Amplifier

The current output of a photodetector is well below the level required to drive an adequate voltage for operating digital logic. The received signal needs to be regenerated with an amplifier before it can be meaningfully used by a circuit. The amplifier component actually consists of two separate types of amplifier circuits. First is a trans-impedance amplifier (TIA) which is responsible for producing a voltage based on an input current. This conversion of current to voltage is necessary because the photo-detector produces a current-based output from the detected photocurrent. A limiting amplifier (LA) can be optionally used at the output of the TIA to increase the peak-to-peak voltage swing of the electrical signal to levels that are suitable for transistors.

The composition of the photo-detector and the amplifier is traditionally known as a receiver.

2.5.4 Deserialization

The deserializer mirrors the functionality of the serializer by down-converting the high-speed optical data to a clock rate that is suitable for the electronic data bus. The output of each receiver typically feeds into a deserializer which produces several parallel data streams at the lower clock rate from a single high-speed data stream. Continuing with the earlier example for the serializer, the deserialization can be used to convert each 16 GHz data stream into 8 unique 2 GHz data streams. A deserializer would be required for each wavelength channel.

2.5.5 Decoder

Lastly, any encoding that was utilized at the generation end of the optical link needs to be reverted so that the pure data signal can be recovered. In the case of the error detection and correction coding, errors need to be addressed appropriately either through error correct or packet retransmission.

References

1. D. Van Thourhout, P. Dumon, W. Bogaerts, G. Roelkens, D. Taillaert, G. Priem, and R. Baets, "Recent progress in SOI nanophotonic waveguides", in *Optical Communication, 2005. ECOC 2005. 31st European Conference on*, vol. 2, Sep. 2005, pp. 241–244.
2. F. Xia, L. Sekaric, and Y. Vlasov, "Ultracompact optical buffers on a silicon chip", *Nature Photonics*, vol. 1, pp. 65–71, 2006.

3. M. Gnan, S. Thorns, D. Macintyre, R. De La Rue, and M. Sorel, "Fabrication of low-loss photonic wires in silicon-on-insulator using hydrogen silsesquioxane electron-beam resist", *Electronics Letters*, vol. 44, no. 2, pp. 115–116, Jan. 2008.
4. J. Cardenas, C. B. Poitras, J. T. Robinson, K. Preston, L. Chen, and M. Lipson, "Low loss etchless silicon photonic waveguides", *OSA Optics Express*, vol. 17, no. 6, pp. 4752–4757, 2009.
5. M. J. Shaw, J. Guo, G. A. Vawter, S. Habermehl, and C. T. Sullivan, "Fabrication techniques for low-loss silicon nitride waveguides", *Micromachining Technology for Micro-Optics and Nano-Optics III*, vol. 5720, no. 1, pp. 109–118, 2005.
6. W. Bogaerts, P. Dumon, D. V. Thourhout, and R. Baets, "Low-loss, low-cross-talk crossings for silicon-on-insulator nanophotonic waveguides", *OSA Optics Letters*, vol. 32, no. 19, pp. 2801–2803, 2007.
7. B. Little, J. Foresi, G. Steinmeyer, E. Thoen, S. Chu, H. Haus, E. Ippen, L. Kimerling, and W. Greene, "Ultra-compact Si-SiO_2 microring resonator optical channel dropping filters", *IEEE Photonics Technology Letters*, vol. 10, no. 4, pp. 549–551, Apr. 1998.
8. B. Lee, A. Biberman, P. Dong, M. Lipson, and K. Bergman, "All-optical comb switch for multiwavelength message routing in silicon photonic networks", *IEEE Photonics Technology Letters*, vol. 20, no. 10, pp. 767–769, May 2008.
9. S. Manipatruni, Q. Xu, B. Schmidt, J. Shakya, and M. Lipson, "High speed carrier injection 18 Gb/s silicon micro-ring electro-optic modulator", in *The 20th Annual Meeting of the IEEE Lasers and Electro-Optics Society (LEOS)*, Oct. 2007, pp. 537–538.
10. H. Lira, S. Manipatruni, and M. Lipson, "Broadband hitless silicon electro-optic switch for optical networks on-chip", in *Group IV Photonics, 2009. GFP '09. 6th IEEE International Conference on*, Sep. 2009, pp. 253–255.
11. Y. Vlasov, W. M. J. Green, and F. Xia, "High-throughput silicon nanophotonic wavelength-insensitive switch for on-chip optical networks", *Nature Photonics*, vol. 2, pp. 242–246, Apr. 2008.
12. Q. Xu, S. Manipatruni, B. Schmidt, J. Shakya, and M. Lipson, "12.5 Gbit/s carrier-injection-based silicon micro-ring silicon modulators", *OSA Optics Express*, vol. 15, no. 2, pp. 430–436, 2007.
13. M. Watts, D. Trotter, R. Young, and A. Lentine, "Ultralow power silicon microdisk modulators and switches", in *5th IEEE International Conference on Group IV Photonics*, Sep. 2008, pp. 4–6.
14. J. Chan, G. Hendry, K. Bergman, and L. Carloni, "Physical-layer modeling and system-level design of chip-scale photonic interconnection networks", *Computer-Aided Design of Integrated Circuits and Systems, IEEE Transactions on*, vol. 30, no. 10, pp. 1507–1520, Oct. 2011.
15. M. Lipson, "Compact electro-optic modulators on a silicon chip", *IEEE Journal of Selected Topics in Quantum Electronics (JSTQE)*, vol. 12, no. 6, pp. 1520–1526, Nov.-Dec. 2006.
16. W. Green, H. Hamann, L. Sekaric, M. Rooks, and Y. Vlasov, "Ultra-compact reconfigurable silicon optical devices using micron-scale localized thermal heating", in *Proceedings of Optical Fiber Communication Conference (OFC)*, Mar. 2007.
17. K. Preston, N. Sherwood-Droz, J. Levy, and M. Lipson, "Performance guidelines for WDM interconnects based on silicon microring resonators", in *Lasers and Electro-Optics (CLEO), 2011 Conference on*, May 2011.
18. J. Feng, Q. Li, and Z. Zhou, "Single ring interferometer configuration with doubled free-spectral range", *Photonics Technology Letters, IEEE*, vol. 23, no. 2, pp. 79–81, Jan. 2011.
19. J. Garcia, A. Martinez, and J. Marti, "Optical add-drop multiplexer with FSR higher than 140 nm using ring resonators and photonic bandgap structures", in *Group IV Photonics, 2008 5th IEEE International Conference on*, Sep. 2008, pp. 82–84.
20. Y. Yanagase, S. Suzuki, Y. Kokubun, and S. T. Chu, "Box-like filter response and expansion of FSR by a vertically triple coupled microring resonator filter", *Lightwave Technology, Journal of*, vol. 20, no. 8, pp. 1525–1529, Aug. 2002.
21. A. X. Widmer and P. A. Franaszek, "A DC-balanced, partitioned-block, 8 b/10 b transmission code", *IBM J. Res. Dev.*, vol. 27, no. 5, pp. 440–451, Sep. 1983.

22. V. Maxwell, "Optical networking", *Potentials, IEEE*, vol. 18, no. 1, pp. 26–28, Feb.-Mar. 1999.
23. H. Sutter, "The Free Lunch Is Over: A Fundamental Turn Toward Concurrency in Software", *Dr. Dobb's Journal*, vol. 30, no. 3, 2005.
24. "Universal serial bus", http://www.usb.org.
25. "The serial ATA international organization", http://www.sata-io.org.
26. C. Severance, "Firewire finally comes home", *Computer*, vol. 31, no. 11, pp. 117–118, Nov. 1998.
27. S. Kimura, T. Hayakawa, T. Horiyama, M. Nakanishi, K. Watanabe, "An on-chip high speed serial communication method based on independent ring oscillators", in *Solid-State Circuits Conference, 2003. Digest of Technical Papers. ISSCC. 2003 IEEE, International*, vol. 1, pp. 390–391, 2003.
28. U. Singh, L. Li, and M. Green, "A 34 Gb/s 2:1 MUX/CMU based on a distributed amplifier using 0.18 μm CMOS", in *VLSI Circuits, 2005. Digest of Technical Papers. 2005 Symposium on*, Jun. 2005, pp. 132–135.
29. A. Rylyakov, C. Schow, B. Lee, W. Green, J. Van Campenhout, M. Yang, F. Doany, S. Assefa, C. Jahnes, J. Kash, and Y. Vlasov, "A 3.9 ns 8.9 mW 4 × 4 silicon photonic switch hybrid integrated with CMOS driver", in *Solid-State Circuits Conference Digest of Technical Papers (ISSCC), 2011 IEEE International*, Feb. 2011, pp. 222–224.
30. N. Ophir, K. Padmaraju, A. Biberman, L. Chen, K. Preston, M. Lipson, and K. Bergman, "First demonstration of error-free operation of a full silicon on-chip photonic link", in *Optical Fiber Communication Conference*, 2011, p. OWZ3.
31. C. Batten, A. Joshi, J. Orcutt, A. Khilo, B. Moss, C. Holzwarth, M. Popovic, H. Li, H. Smith, J. Hoyt, F. Kartner, R. Ram, V. Stojanovic, and K. Asanovic, "Building many-core processor-to-DRAM networks with monolithic CMOS silicon photonics", *IEEE Micro*, vol. 29, no. 4, pp. 8–21, Jul.-Aug. 2009.
32. N. Kirman, M. Kirman, R. K. Dokania, J. F. Martinez, A. B. Apsel, M. A. Watkins, and D. H. Albonesi, "On-chip optical technology in future bus-based multicore designs", *IEEE Micro*, vol. 27, no. 1, pp. 56–66, 2007.
33. M. J. Cianchetti, J. C. Kerekes, and D. H. Albonesi, "Phastlane: a rapid transit optical routing network", in *Proceedings of the 36th Annual International Symposium on Computer Architecture (ISCA)*, 2009, pp. 441–450.
34. D. Vantrease, R. Schreiber, M. Monchiero, M. McLaren, N. P. Jouppi, M. Fiorentino, A. Davis, N. Binkert, R. G. Beausoleil, and J. H. Ahn, "Corona: System implications of emerging nanophotonic technology", *Proceedings of the 35th Annual International Symposium on Computer Architecture (ISCA)*, vol. 0, pp. 153–164, Jun. 2008.
35. A. Shacham, K. Bergman, and L. Carloni, "Photonic networks-on-chip for future generations of chip multiprocessors", *IEEE Transactions on Computers*, vol. 57, no. 9, pp. 1246–1260, Sep. 2008.
36. G. Hendry, S. Kamil, A. Biberman, J. Chan, B. Lee, M. Mohiyuddin, A. Jain, K. Bergman, L. Carloni, J. Kubiatowicz, L. Oliker, and J. Shalf, "Analysis of photonic networks for a chip multiprocessor using scientific applications", in *Proceedings of the 2009 3rd ACM/IEEE International Symposium on Networks-on-Chip (NOCS)*, May 2009, pp. 104–113.
37. J. Chan, G. Hendry, A. Biberman, and K. Bergman, "Architectural exploration of chip-scale photonic interconnection network designs using physical-layer analysis", *IEEE/OSA Journal of Lightwave Technology*, vol. 28, no. 9, pp. 1305–1315, May 2010.
38. G. Hendry, J. Chan, S. Kamil, L. Oliker, J. Shalf, L. Carloni, and K. Bergman, "Silicon nanophotonic network-on-chip using TDM arbitration", in *2010 IEEE 18th Annual Symposium on High Performance Interconnects (HOTI)*, Aug. 2010, pp. 88–95.
39. J. Chan and K. Bergman, "Photonic interconnection network architectures using wavelength-selective spatial routing for chip-scale communications", *J. Opt. Commun. Netw.*, vol. 4, no. 3, pp. 189–201, Mar. 2012.

Chapter 3
Silicon Photonics

Systems harnessing silicon photonic technology have the potential to vastly improve the performance of interconnects and computing systems. This chapter provides insight into this emerging technology, and summarizes several critical empirical results. Key conclusions that can be drawn from these demonstrations are highlighted, followed by discussions on relevant performance metrics and their corresponding challenges.

3.1 Materials

Silicon (Si) can be considered the workhorse of the semiconductor industry. Given its abundance and versatility, silicon has become the dominant platform for microelectronic fabrication. As such, any improvements to CMOS microelectronic chips must be compatible with the silicon materials, devices, and processing. Below is a discussion of the main CMOS materials with potential applications for integrated photonics.

3.1.1 Crystalline Silicon

Single-crystalline silicon (c-Si) consists of a perfect lattice of silicon atoms, and offers the best set of electrical and optical properties for silicon integrated photonics. Silicon is transparent at the standard telecommunication wavelengths around $\lambda = 1.55\,\mu\text{m}$, and has a high refractive index of 3.5 that allows sub-micron-scale waveguides with very tight bending radii when paired with a silicon dioxide (SiO_2) cladding. The fundamental optical losses in bulk crystalline silicon at $\lambda = 1.55\,\mu\text{m}$ are dominated by free-carrier absorption [1]. Electrons and holes also induce a small change in the refractive index of silicon, which can be harnessed as an efficient and effective way

K. Bergman et al., *Photonic Network-on-Chip Design*,
Integrated Circuits and Systems, DOI: 10.1007/978-1-4419-9335-9_3,

to modulate and route light on chip [2, 3]. Typical p-doped silicon waveguides have a concentration of 10^{15} cm^{-3} holes, generating a fundamental material propagation loss of 0.026 dB/cm from this effect.

Experimentally, the waveguide propagation loss is higher due to scattering and absorption at the waveguide sidewalls. Strip, locally-oxidized, and ridge silicon waveguides have achieved propagation losses of 80 [4], 21 [5], and about 2.6–20 dB/m [6–9], respectively. Silicon microdisk resonators have been shown to have propagation losses as low as 10 dB/m, with quality factors of $Q = 5{,}000{,}000$, limited by absorption in surface states [8]. Other research efforts have leveraged reflowing-photoresist and oxidation-based-smoothing techniques to produce ridge waveguides that substantially reduce sidewall interaction. This yielded silicon ring resonators with measured propagation losses of 2.7 dB/m, and internal quality factors of $Q = 22{,}000{,}000$ [10, 11]. It is expected that future waveguide demonstrations will achieve even lower bounds in propagation loss.

Unfortunately crystalline silicon can only be grown from another silicon crystal, making it impossible to deposit. This restriction typically limits the optical devices to a single layer. Most research in the area of silicon photonics thus relies on a custom silicon wafer not typically used in standard CMOS foundries. The wafer is known as silicon on insulator (SOI) and is composed of a ~0.25 μm crystalline silicon device layer on top of a thick buried silicon dioxide optical buffering layer on top of a bulk silicon substrate. The SOI process typically raises the cost of a wafer by at least an order of magnitude, and limits the architecture to a single optical layer of waveguides and other devices.

3.1.2 Polycrystalline Silicon

Polycrystalline silicon (polysilicon, or poly-Si) is commonly found in CMOS fabrication as a gate material for transistors. However, it is also capable of being deposited as an electrically-conductive light-guiding material. Polycrystalline silicon does not have a homogeneous crystalline structure, but instead consists of crystalline grains separated by thin disordered grain boundaries. The size of the crystalline grains and the nature of the grain boundaries have a significant effect on the optical and electrical material properties, and are controllable by the fabrication conditions. Optically, the grain boundaries scatter and absorb light and cause propagation losses, attenuating the optical signal as it propagates through the material. Electrically, the grain boundaries impede the flow of electrons and holes. Hence, having larger grains (fewer grain boundaries) in the material optimizes both the optical and electrical properties. Experimentally, the minimum demonstrated waveguide propagation loss is 6.45 dB/cm [12]. Innovations in fabrication techniques will enable experimental demonstrated with significantly lower loss properties.

Although for many applications the electrical properties of polycrystalline silicon are inferior to crystalline silicon, the grain boundaries reduce the free-carrier lifetime, which can be advantageous for some devices. The effective free-carrier recombination time has been demonstrated to be $\tau_{fc} \approx 100$ ps [13], or approximately four times faster than comparable devices in crystalline silicon. Additionally, it has been shown that polycrystalline silicon can be used as a moderately absorbing material for photodetectors [14].

3.1.3 Silicon Nitride

Although polycrystalline silicon may be used for electrically active devices, the propagation losses are intolerable over centimeter-scale distances. Silicon nitride (Si_3N_4) is another CMOS-compatible material capable of being deposited, with a refractive index near 2 at $\lambda = 1.55\,\mu m$. Microelectronic processing typically uses silicon nitride as a masking layer. Although silicon nitride is not useful for making electrically-active devices, its high refractive index, compared to silicon dioxide, still allows for high confinement sub-micrometer waveguides.

Very weakly confined modes in silicon nitride waveguides reveal that losses as low as 0.045 dB/m are possible [15]. Highly-confined silicon nitride waveguides have achieved about 10 dB/m [16, 17]. This represents approximately an order of magnitude lower propagation loss compared to crystalline silicon waveguides, making it an ideal candidate for waveguides in a deposited material.

3.1.4 Amorphous Silicon

Hydrogenated amorphous silicon (a-Si:H) is another deposited material that can be used for waveguides. The refractive index (3.5) and waveguide propagation loss (2–3 dB/cm) are comparable to crystalline silicon, but the propagation loss can increase dramatically with time and/or temperature if hydrogen diffuses out [18]. Without a strictly controlled environment, a-Si:H is a much less stable material than silicon nitride.

3.1.5 Germanium

Photodetectors require a material that absorbs light at the wavelength of interest. Germanium (Ge) is an available CMOS material used to create strained silicon transistors and SiGe alloy heterojunction bipolar transistors. Germanium is an indirect-bandgap material whose 0.8-eV direct-band transition enables strong absorption for $\lambda \leq 1.55\,\mu m$, and strain can be used to extend absorption to longer wavelengths. The primary challenge for germanium integration is the 4 % lattice mismatch with silicon, which makes it difficult to grow high-quality germanium structure [19].

Polycrystalline germanium can also be used as a lower-quality photodetector material that is easier to integrate [20].

3.1.6 Silicon Dioxide

Silicon dioxide (SiO_2 or silica) is a material that exhibits good electrical insulation and high thermal stability. For this reason, thin films of silicon dioxide grown on silicon wafers are used in microelectronics to achieve electric insulation with high chemical stability. In photonic integration, the relatively-low refractive index of about 1.5 near $\lambda = 1.55\,\mu m$ makes silicon dioxide a good cladding material choice for high-index-contrast material systems. This enables high confinement of light inside the high-refractive-index core materials, and subsequently dense integration of photonic components on chip.

Because silicon dioxide can have extremely low optical absorption and scattering losses, there have also been demonstrations using silicon dioxide as the core material for waveguides with low propagation losses. Silicon dioxide waveguides have achieved about 0.037 dB/m [21], and silicon dioxide wedge resonators have achieved quality factors of $Q = 875{,}000{,}000$ [22]. Phosphorus-doped silica-on-silicon waveguides have achieved 0.85 dB/m [23].

The majority of optical fibers utilized in telecommunications networks are also made from silicon dioxide. These optical fibers are cylindrical dielectric waveguides, which confine and transmit light through the process of total internal reflection. Similar to any other type of waveguide, the optical fiber consists of a high refractive-index core material surrounded by a lower-refractive-index cladding layer. Silicon dioxide may be doped with various materials to raise or lower its refractive index.

3.1.7 Material Stacks

An optical layer consists of a high-refractive-index core material sandwiched between lower-index cladding layers. The cladding provides a buffer to keep the light confined to the waveguide core. An example optical layer is depicted in Fig. 3.1, which leverages the SOI platform to produce both passive and active optical functionalities using a single layer of crystalline silicon, along with germanium for photodetectors.

Multiple optical layers can be stacked with a buffer layer of cladding material in between them. The thickness of the buffer controls the optical coupling between the layers in the same way a horizontal gap is used between a single-layer waveguide coupling to a microring resonator. One benefit of multi-layer integration is that deposition of these materials can be controlled accurately to the nanometer scale, which is prohibitively difficult and expensive for lithographically-defined single-layer gaps.

Another example material stack is depicted in Fig. 3.2, containing three optical layers: an active polycrystalline silicon layer sandwiched between two silicon

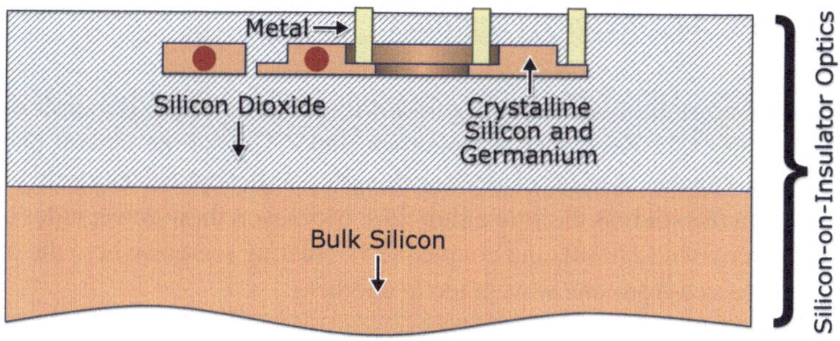

Fig. 3.1 Cross-Sectional View of Silicon-on-Insulator (SOI) Photonic Stack: Cross-sectional view of crystalline silicon optical layer. Crystalline silicon bus waveguide coupled to crystalline silicon microring resonator

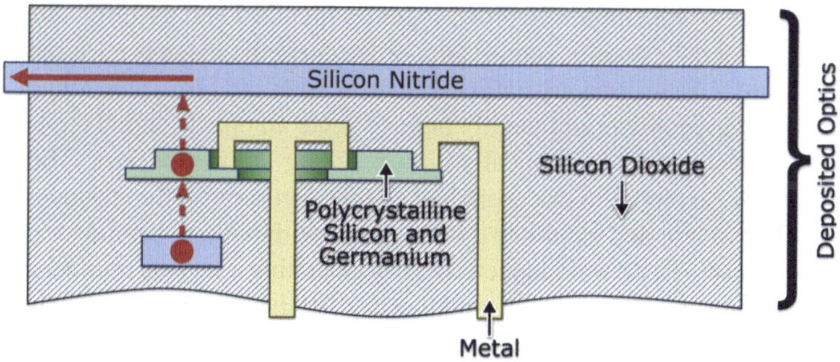

Fig. 3.2 Cross-Sectional View of 3D Silicon Photonic Stack [24]: Cross-sectional view of vertically-stacked optical link. Two orthogonal silicon nitride bus waveguides are shown above and below of an electrically-active region which consists of polycrystalline silicon or germanium

nitride waveguiding layers, with silicon dioxide cladding in between the layers. This approach provides an active layer (used for switching, photodetection, and modulation) coupled to two bus waveguide layers which do not interact with each other. There is no fundamental limitation on how many layers one could use, given an architectural need for them, since each layer is deposited and each buffer layer is flattened using chemical-mechanical planarization.

3.2 Waveguides

The waveguide is the most basic silicon photonic component, used to carry high-speed optical data from one point to another [25, 26]. In most silicon photonic applications, the high index contrast between the silicon waveguides and the silicon dioxide cladding surrounding them results in an extremely small optical mode size for single-mode operation. This condition enables very dense integration of silicon photonic devices.

3.2.1 Crystalline Silicon Waveguides

In recent years, crystalline silicon waveguides with sub-micrometer dimensions have been almost the exclusive choice for photonic links. Crystalline silicon photonic waveguides are capable of transporting wavelength-parallel optical data with terabit-per-second data rates across the entire chip [26]. Moreover, these waveguides can be bent [26, 27], crossed [27–30], and coupled [25], creating regions where the optical signal can be passed from one waveguide to another.

3.2.1.1 1.28-Tb/s Optical Data Stream Through Crystalline Silicon Waveguide

One of the first experimental demonstrations of transmitting high-speed optical data through these crystalline silicon waveguides included successfully launching a 1.28-Tb/s optical data stream, comprising 32 wavelength channels each modulated at 40 Gb/s, into a 5-cm-long crystalline silicon waveguide [26]. This length of the demonstrated waveguide is sufficiently large to route an optical signal anywhere on chip, with a given chip edge typically not exceeding 2.5 centimeters. The single-mode waveguide explored in that work had a cross section with dimensions 520-nm wide and 220-nm tall, and exhibited 24 90°-bends with 6.5-μm bending radii.

Inter-channel crosstalk was characterized by launching a varying number of wavelength channels from one to 24, each modulated at 10 Gb/s, and quantifying the resulting signal degradation on one of the wavelength channels. Furthermore, the effects of self-phase modulation (SPM) were quantified in these waveguides by launching a single wavelength channel modulated at 10 Gb/s, and varying the input peak optical power.

These experiments demonstrate that these crystalline silicon waveguides:

- Have the ability to successfully transmit high-speed optical data, at least as high as 40 Gb/s per wavelength channel, anywhere on chip.
- Can contain many bends without adding significant signal degradation, enabling the ability to construct complex photonic network topologies.
- Exhibit an enormous amount of usable bandwidth that can be leveraged to transmit a massive amount of optical data by scaling the data rate or increasing the wavelength parallelism.
- Support wavelength-parallel transmission scalable to multi-terabit-per-second aggregate data rates.

3.2.1.2 Crystalline Silicon Waveguide Crossings

Waveguide crossings are generally an unavoidable part of every integrated photonic interconnection network. One significant limitation of crystalline silicon waveguides, common to all high-index-contrast material systems, is that physical crossings of waveguides result in significant insertion losses when considering a full-scale interconnection network [31]. These insertion losses come from the fact that any abrupt directional changes in dielectric waveguides causes mode conversion into unguided modes, translating to radiation losses [27]. Ideally, the optical signal at a waveguide crossing is transmitted forward without optical power escaping sideways into other waveguides (in form of crosstalk), reflecting back at the interface (in form of back-reflection), or dissipating into the substrate.

Significant research efforts have been employed to minimize these losses. Using waveguide broadening to expand the waveguide mode, combined with a double-etch technique to locally reduce the refractive index contrast, waveguide crossing insertion loss as low as 0.16 dB/crossing has been experimentally demonstrated, with crosstalk below −40 dB [28]. Using adiabatic tapers to couple the fundamental mode to two lowest-order even modes, combined with the multi-mode interference (MMI) method, waveguide crossing insertion loss as low as 0.12 dB/crossing was proposed and simulated, with crosstalk below −40 dB, and back-reflection about −45 dB [29]. Finally, using short non-adiabatic tapers to excite the appropriate ratio of the fundamental and third mode, combined with periodically-matched multi-mode focusing, waveguide crossing insertion loss as low as 0.045 dB/crossing was proposed and simulated, with crosstalk between −35 and −40 dB [30].

Even with a 0.045-dB/crossing insertion loss, crystalline silicon waveguide crossings add significant power limitations when multiplied by tens, hundreds, or even thousands of instances. This is potentially a major limitation for the complexity and scalability of integrated photonic interconnection networks.

3.2.2 Silicon Nitride Waveguides and Waveguide Crossings

In contrast to crystalline silicon waveguides, deposited silicon nitride waveguides introduce several potential advantages for integrated photonics, many of which are yet to be explored. Recent efforts have placed silicon nitride waveguides as the carrier medium in high-speed communication links, with the vision of monolithic integration of high-performance photonic and electrical elements using standard CMOS processes [32]. Unlike crystalline silicon, silicon nitride can be deposited in multiple layers, analogous to electronic wiring. This multi-layer integration has the potential to eliminate in-plane waveguide crossing losses, as waveguide crossings may now be performed entirely out of plane.

3.2.2.1 1.28-Tb/s Optical Data Stream Through Silicon Nitride Waveguide

To experimentally demonstrate the transmission of high-speed optical data through deposited silicon nitride waveguides, a 1.28-Tb/s optical data stream, comprising 32 wavelength channels each modulated at 40 Gb/s, was successfully launched into a 4.3-cm-long deposited silicon nitride waveguide [33]. Here, the single-mode waveguide was 1,800-nm wide and 750-nm tall. To characterize the inter-channel crosstalk, the number of wavelength channels was varied from one to 32, each modulated at 40 Gb/s.

These experiments demonstrate that the silicon nitride waveguides have all the advantages of crystalline silicon waveguides, as well as:

- Have the ability to be deposited, enabling the deposition of high-quality silicon photonic waveguides in both front- and back-end processes.
- Exhibit no inter-channel crosstalk dependence due to the severely reduced susceptibility to nonlinear effects such as two-photon absorption (TPA), and the resulting free carrier absorption (FCA) and free carrier dispersion (FCD) [17, 34].

3.2.3 Inter-Channel Crosstalk in Crystalline Silicon and Silicon Nitride Waveguides

Figure 3.3 summarizes the inter-channel crosstalk experimental results from Sects. 3.2.1.1 and 3.2.2.1, comparing the power penalty for crystalline silicon [26] and silicon nitride [33] waveguides, respectively [35]. For a signal consisting of one wavelength channel, the power penalty is low and similar for the two material systems. As the number of utilized wavelength channels is increased, the power penalty for the crystalline silicon waveguide increases dramatically, while remaining constant for the silicon nitride waveguide all the way up to 32 wavelength channels. As mentioned in Sect. 3.2.2.1, the silicon nitride waveguide exhibits this low inter-channel crosstalk dependence due to the severely reduced susceptibility to nonlinear effects in this material system, allowing the scaling of the number of wavelength channels in a waveguide without degrading the signal integrity of other wavelength channels in the optical data stream.

3.3 Microring Resonators

Many silicon photonic devices leverage the microring resonator to achieve high-performance operation. Many physical phenomena being explored in silicon photonics are enhanced by these types of resonator cavities, leading to much more compact and efficient silicon photonic devices. The high refractive index contrast in these devices further reduces the necessary size of these microring resonators.

In the passive state, the microring resonator may act as an efficient wavelength-selective switch, filter, multiplexer, and demultiplexer, predetermined to add or drop a target wavelength channel to or from a wavelength-parallel optical data stream. In the active state, the microring resonator may act as an efficient modulator or switch, selectively directing an optical signal to its target destination. The microring resonator may simultaneously behave as a passive wavelength-selective switch and

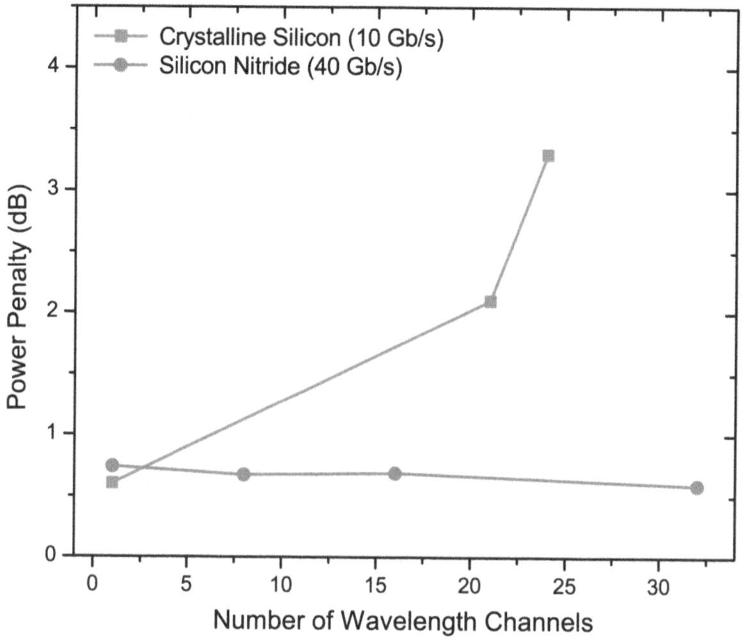

Fig. 3.3 Power Penalty for Waveguide Inter-Channel Crosstalk: Inter-channel crosstalk experimental results from Sects. 3.2.1.1 and 3.2.2.1, comparing the power penalty for crystalline silicon [26] and silicon nitride [33] waveguides, respectively

an active photodetector, selecting a single wavelength channel from a wavelength-parallel optical data stream to be detected.

3.4 Modulators

The electro-optic modulator is a critical device that enables high-speed conversion from an electrical signal to an optical signal. This device typically encodes data on a single wavelength channel that is combined with other optical signals through WDM, forming a cohesive wavelength-parallel optical signal.

3.4.1 Crystalline Silicon Modulators

Due to their compact size, high speed, and low energy dissipation, the crystalline silicon microring resonator electro-optic modulators have been the focus of most recent efforts proposing chip-scale photonic interconnection networks [36–41]. Recent experimental validations have produced these devices with microring resonator diameters as low as 3 μm [42], modulation rates as high as 44 Gb/s, driving voltages as low as 150 mV$_{PP}$ [43], and in the microdisk resonator configuration, energy dissipation as low as 3 fJ/bit [44]. No doubt these metrics will continue to improve with further innovation and refinement.

Crystalline silicon microring modulators, first demonstrated in the injection-based lateral-junction configuration [45], have been primarily considered for these applications because they are compact, high-speed, low-power, and are capable of enabling WDM systems with many wavelength channels. Depletion-based devices improve on these early injection-based device designs, enabling efficient high-speed operation of silicon microring and microdisk modulators. Recent work has yielded depletion-based modulators leveraging lateral-junction configurations capable of operating beyond 25 Gb/s [46–48], and even 40 Gb/s [49–51]. Methods to interleave PN junctions have been utilized to increase the electro-optic efficiency by enhancing the carrier-light overlap, which reduces the voltage-length product $V_\pi L_\pi$, achieving 25-Gb/s operation of a 30-μm-radius device with 4.5-dB extinction ratio at 2-V driving voltage [46], and 25-Gb/s operation of a 250-μm-circumference device with about 3.6-dB extinction ratio at 1.6-V driving voltage and 41-fJ/bit power consumption [47]. The extinction ratio is defined as the ratio between the one and the zero of the optical signal, relative to the true zero. Another demonstration has achieved 25-Gb/s operation of a 7.5-μm-radius device with about 5-dB extinction ratio at 1-V driving voltage and about 7-fJ/bit power consumption [48]. By reducing the radius of the latter device to 5 μm and reducing the junction capacitance and thus the RC time constant (the time required to charge or discharge the capacitor of the device, through the resistor of the device), a similar device was demonstrated to operate at 40 (44) Gb/s, with 7- (2.3–) dB extinction ratios at 2-V driving voltages [49]. Leveraging PN-junction-interleaved cascaded microring resonators, 40-Gb/s operation was demonstrated with a 20-μm-radius device with 3.9-dB extinction ratio at 5-V driving voltage [50]. Lastly, a method to zigzag the PN junction has been employed to achieve 44-Gb/s operation with 3.01-dB extinction ratio at 3-V driving voltage [51].

Depletion-based devices leveraging vertical-junction configurations provide the added advantage to the drive voltage, power consumption, size, and speed [52]. The vertical junction naturally maximizes the overlap of the depletion region with the optical mode, reducing the drive voltage and power consumption. It also enables direct electrical contacts to be made at the interior of the silicon microdisk or microring resonator, forming a hard outer wall that minimizes radiation, resulting in a significant size reduction of the device. This reduction in size also reduces the capacitance, thereby reducing the RC time constant and increasing the speed. For example, depletion-based vertical-junction silicon microdisk modulators have enabled 12.5-Gb/s modulation of a 1.75-μm-radius device with 3.2-dB extinction ratio at 1-V driving voltage and 3-fJ/bit power consumption [52]. This device occupies less than one-tenth the area of traditional microring modulators, and maintains entirely-interior electrical contacts.

However, one major limitation of microdisk resonators is that they support higher-order spatial modes that corrupt the FSR by introducing undesired resonances.

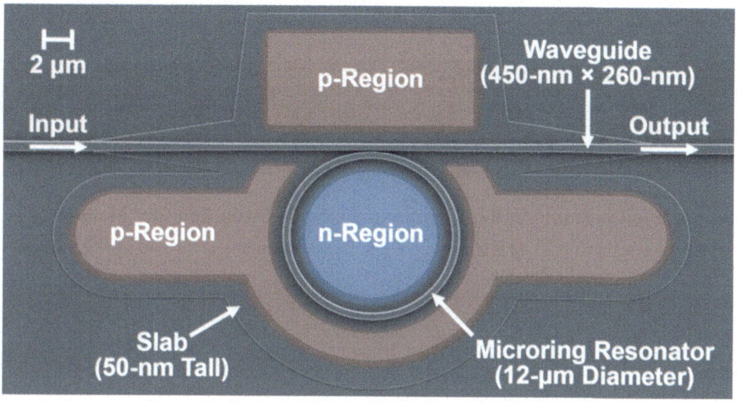

Fig. 3.4 Crystalline Silicon Microring Resonator Electro-Optic Modulator [59]: *Top-view* scanning-electron-microscope (SEM) image of the crystalline silicon microring resonator electro-optic modulator

To combat this, several demonstrations have utilized adiabatic microring resonators, which leverage single-mode coupling regions and multimode contact regions to achieve single-mode operation [53–56]. These devices preserve all the benefits of microdisk resonators, with the added advantage of enabling uncorrupted FSRs [55]. One demonstration has directly integrated heaters within adiabatic microring resonators, achieving 4.4-μW/GHz power dissipation and 1-μs thermal tuning speed of a second-order filter [56]. Another demonstration has presented preliminary results of a depletion-based vertical-junction adiabatic microring modulator, with an uncorrupted 6.9-THz FSR, achieving 12-Gb/s operation of a 2-μm-radius device with about 7.3-dB extinction ratio at 2.4-V driving voltage [53, 54].

3.4.1.1 Comparative Performance Analysis of Crystalline Silicon Microring Resonator Modulator with Modulation Rate up to 12.5 Gb/s

One demonstrated crystalline silicon microring resonator electro-optic modulator comprised a microring resonator configured as a P-I-N carrier injection device [57–59]. This microring resonator has a 12-μm diameter, optical quality factor of about 10,000, and is coupled to a straight waveguide with input and output ports, as depicted in Fig. 3.4. The waveguides were 450-nm wide and 260-nm tall. The waveguide of the microring has a 50-nm slab that was doped to form the P-I-N diode structure, with nickel silicide for the electrical contacts. The crystalline silicon modulator was fabricated using electron beam lithography (EBL) and reactive ion etching (RIE) [60].

During standard operation of this device, a non-return to zero (NRZ) on-off keyed (OOK) modulation signal is electro-optically encoded onto a wavelength channel when an incoming continuous wave (CW) source passes in and out of a shifting

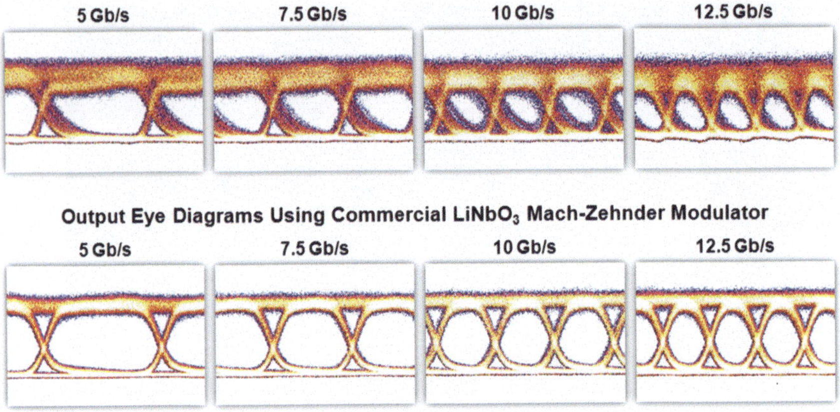

Fig. 3.5 Output Eye Diagrams Using Microring Resonator Modulator [58]: Experimentally-measured temporal responses of varying modulation rates for the silicon microring resonator electro-optic modulator as well as a commercial $LiNbO_3$ electro-optic MZM. Output eye diagrams for the 5-, 7.5-, 10-, and 12.5-Gb/s modulation rates using the silicon modulator, and 5-, 7.5-, 10-, 12.5-Gb/s modulation rates using the $LiNbO_3$ modulator

resonance of the microring resonator. When the resonance of the microring resonator is tuned onto the wavelength of the incoming signal, the light is coupled into the microring resonator, corresponding to a logical *0* bit. When the resonance of the microring resonator is tuned away from the wavelength of the incoming signal, the signal bypasses the microring resonator and leaves on the output port, corresponding to a logical *1* bit. The resonance shift is induced by the plasma-dispersion effect from injecting and extracting electrical carriers through the PIN diode, which is integrated into the microring resonator. To achieve high modulation rates that are typically limited by carrier lifetimes, the modulator may be driven using a pre-emphasis method [2].

This device has been demonstrated to achieve feasible operation for modulation rates up to 12.5 Gb/s, and a comparison of this device was performed with a commercial $LiNbO_3$ electro-optic MZM [57, 58].

The silicon modulator was first evaluated at varying modulation rates by examining the output modulation temporal response. This was accomplished by electrically driving the silicon modulator with 5-, 7.5-, 10-, and 12.5-Gb/s NRZ OOK data. The electro-optic response of the silicon modulator then encoded the incoming light with the electrical data. After the data signal has left the chip, the eye diagrams of the optical signals were measured, as depicted in Fig. 3.5. The resulting eye diagrams show clear openings, and degrade as the modulator is driven at higher modulation rates, resulting from electrical carrier lifetime limitations as well as the transient response of the microring resonator and the closing of the temporal window.

The performance of the silicon modulator was then compared with a commercial $LiNbO_3$ electro-optic Mach-Zehnder modulator (MZM) at varying modulation rates. This was accomplished by setting the clock rate to 5, 7.5, 10, and 12.5 GHz,

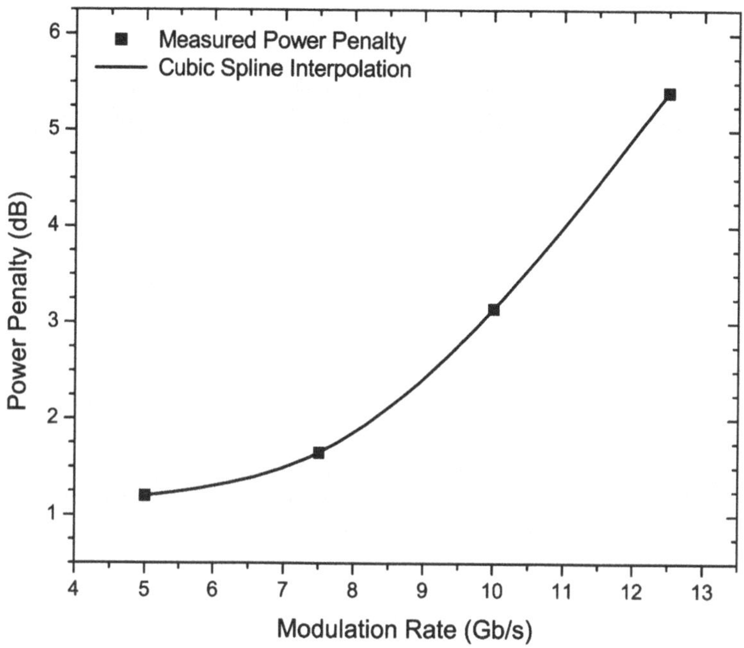

Fig. 3.6 Power Penalty Using Silicon Microring Resonator Modulator Compared to Commercial $LiNbO_3$ Mach-Zehnder Modulator [58]: Resulting power penalty associated with the operation of the silicon modulator compared to the $LiNbO_3$ modulator for modulation rates between 5 and 12.5 Gb/s

electrically driving the $LiNbO_3$ modulator with 5-, 7.5-, 10-, and 12.5-Gb/s data. The electro-optic response of the $LiNbO_3$ modulator then encoded the incoming light with this data. The eye diagrams of the optical signal were evaluated, as depicted in Fig. 3.5. The resulting eye diagrams show clear openings, and degrade as the modulator is driven at higher modulation rates, resulting mostly from the closing of the temporal window.

To draw a system-level functional comparison between the two modulators, the resulting power penalties of the operation of the silicon modulator compared to the $LiNbO_3$ modulator are depicted in Fig. 3.6. These power penalties may be further improved with more optimal pre-emphasis configurations for each modulation rate, as well as closer integration of the electrical driving circuit with the silicon modulator.

These experiments demonstrate that the microring resonator electro-optic modulators:

- Possess a flexible range of modulation rates, with error-free operation up to 12.5 Gb/s. Feasible data rates will continue to increase as these devices are optimized by the research community.
- Already have comparable performance to commercial alternatives, especially for lower modulation rates. Further optimizations can be applied to the silicon

modulator, such as developing a more efficient pre-emphasized electrical driving signal [2], to close (and even potentially exceed) this performance gap.
- Offer a direct trade-off between modulation rate and signal integrity, which can be optimized to maximize system-level performance for any integrated photonic interconnection network.

3.4.1.2 Long-Haul Transmission Using Crystalline Silicon Microring Resonator Modulator with Bandwidth-Distance Product up to 1 Tb-km/s

Silicon photonics is slated to deliver potentially orders-of-magnitude performance improvements in both bandwidth and power consumption by interconnecting chip-scale high-performance computing systems and memory access networks, as well as augmenting data center interconnection networks. Until recently, the functionalities of silicon photonics for medium- and long-haul optical communications remained largely unexplored. To close this gap, a feasible long-haul (up to 80-km) transmission was demonstrated using the high-speed (up to 12.5-Gb/s) and compact (12-μm diameter) silicon microring modulator described in Sect. 3.4.1.1, showing a bandwidth-distance product up to 1,000 Gb-km/s [57, 59].

The signal integrity degradation induced by the chromatic dispersion effects caused by varying propagation distances through standard single mode fiber (SSMF) was studied. This degradation results from the induced chirp in the silicon modulator. A 10-Gb/s modulated signal was first transmitted through SSMF lengths of 0, 1, 2, 5, 10, 15, 40, 60, and 80 km, inducing proportional amounts of chromatic dispersion. The eye diagrams of the resulting optical signals were evaluated for each configuration, as depicted in Fig. 3.7. The eye diagrams show clear openings, and remain relatively unchanged for optical fiber lengths up to 15 km. The eye diagrams subsequently began to lose temporal window and display increased noise after 40-km propagation distances, as dispersion effects become more distinct. Increasingly noticeable degradation occurred at 60 km and 80 km. To evaluate this dependence of the signal integrity degradation on modulation rate, the silicon modulator was subsequently evaluated at a 12.5-Gb/s modulation rate for the 0- and 80-km optical fiber lengths, as depicted in Fig. 3.7. After this 80-km transmission, the eye diagram of the resulting optical signal was evaluated. Again, the eye diagram showed noticeable degradation from the induced chromatic dispersion.

The signal integrity degradation caused by the chromatic dispersion effects induced by long-haul propagation was quantified at varying propagation distances through SSMF. The measured power penalty was recorded for each propagation distance, as shown in Fig. 3.8.

Using the chromatic dispersion for a given propagation distance characteristic of the SSMF employed in these measurements, which is about 17 ps/(nm-km) at the 1565-nm operating wavelength, the measured power penalty for each bandwidth-distance product configuration were then determined. These experimental results are summarized in Fig. 3.9.

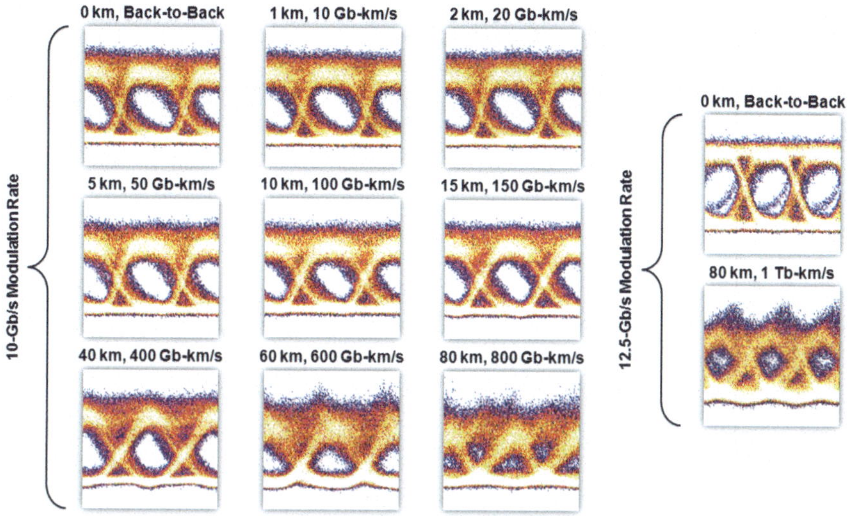

Fig. 3.7 Output Eye Diagrams Using Silicon Microring Resonator Modulator and Standard Single-Mode Optical Fiber [59]: Experimentally-measured temporal responses of varying propagation distances through standard single-mode optical fiber of an optical signal encoded using the silicon microring resonator electro-optic modulator. Output eye diagrams for the 0-, 1-, 2-, 5-, 10-, 15-, 40-, 60-, and 80-km propagation distances for 10-Gb/s modulation rates, as well as 0- and 80-km propagation distances for 12.5-Gb/s modulation rate, using the silicon modulator

Quantifiable performance metrics extracted from experimental validation of silicon photonic devices aid in determining the functionality that these devices perform in large-scale photonic network architectures. These experiments demonstrate that the silicon microring resonator electro-optic modulators also:

- Are truly versatile silicon photonic devices, capable of enabling high-performance transmission for a wide range of short-, medium-, and long-haul applications.
- Offer high-quality transmission in potential applications ranging from chip-scale high-performance computing systems, memory access networks, data center interconnection networks, and even telecommunication networks.

3.4.2 Crystalline Silicon Microring Resonator Electro-Optic Modulator Arrays

To maximize the total transmitted bandwidth, integrated photonic interconnection networks employ arrays of modulators, leveraging wavelength parallelism. As depicted in Fig. 3.10, each modulator in the array corresponds to a specific wavelength channel. At the input side of the modulator array, wavelength-parallel CW light is injected, corresponding to each wavelength channel utilized by the photonic

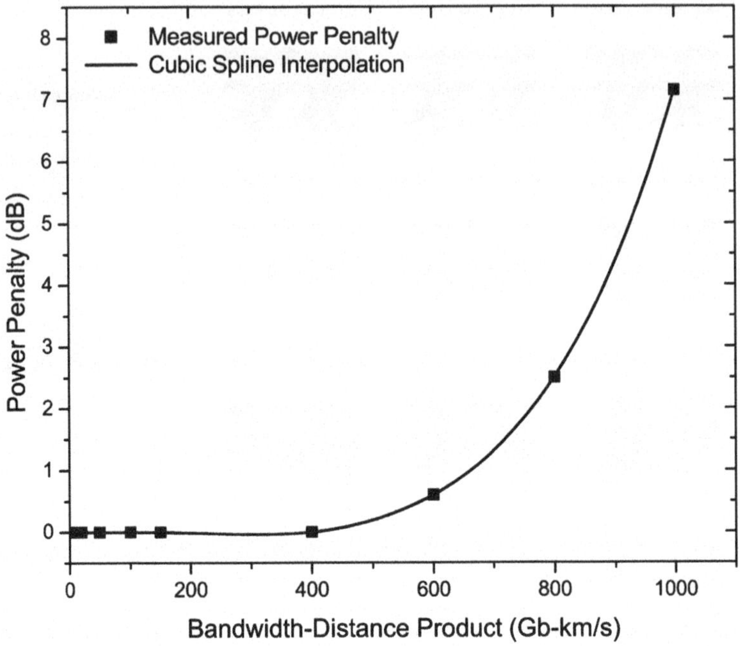

Fig. 3.8 Power Penalty Using Silicon Microring Resonator Modulator and Standard Single-Mode Optical Fiber [59]: Resulting power penalty associated with propagating the optical signal encoded using the silicon modulator through varying lengths of standard single-mode optical fiber for 10-Gb/s modulation rate

Propagation Distance (km)	Modulation Rate (Gb/s)	Induced Chromatic Dispersion (ps/nm)	Measured Power Penalty (dB)	Bandwidth-Distance Product (Gb-km/s)
1	10	17	0	10
2	10	34	0	20
5	10	85	0	50
10	10	170	0	100
15	10	255	0	150
40	10	680	0	400
60	10	1020	0.6	600
80	10	1360	2.5	800
80	12.5	1360	2.5	1000

Fig. 3.9 Summary of Results Using Silicon Microring Resonator Modulator and Standard Single-Mode Optical Fiber [57]: Summary of experimentally-measured results for varying propagation distances through standard single-mode optical fiber, induced chromatic dispersions, and bandwidth-distance products, of an optical signal encoded using the using the silicon microring resonator electro-optic modulator for modulation rates between 10 and 12.5 Gb/s

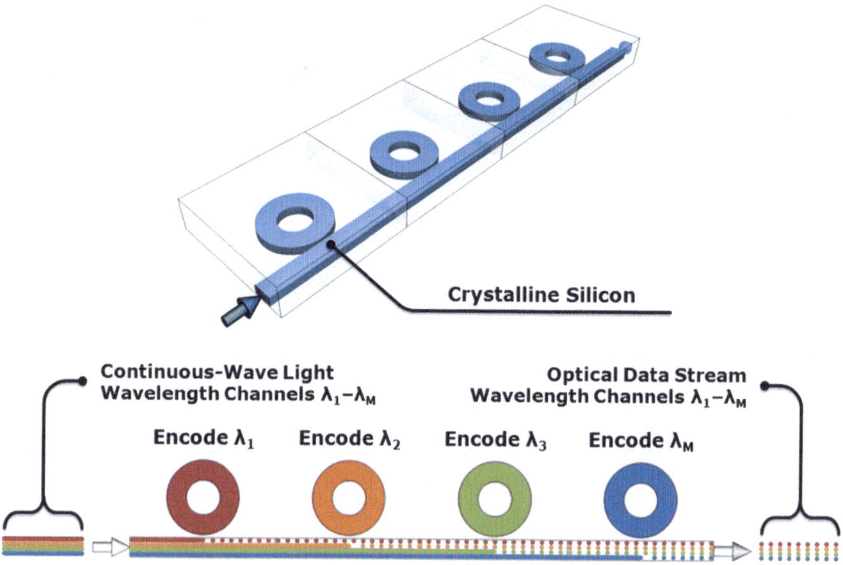

Fig. 3.10 Crystalline Silicon Microring Resonator Modulator Array: Crystalline silicon microring resonator electro-optic modulator array formed by concatenating multiple modulators in series. Each modulator in the array corresponds to a specific wavelength channel. Wavelength-parallel continuous-wave light is injected at the input. Each wavelength channel is selectively encoded with data by a corresponding modulator. Resulting wavelength-parallel optical data stream is extracted at the output

transmitter. As this light propagates through the modulator array, each wavelength channel is selectively encoded with data by a corresponding modulator—each modulator is tuned to modulate one target wavelength channel. At the output side of the modulator array, the resulting wavelength-parallel optical data stream is extracted, which is then ready to be routed through the integrated photonic interconnection network.

One experimental demonstration showing the potential feasibility of the silicon microring resonator electro-optic modulator array individually tested four modulators arranged in an array configuration [61], similar to the one depicted in Fig. 3.10. In this experiment, each modulator was consecutively driven with a 4-Gb/s electrical signal. The modulators were tested for inter-channel crosstalk, which was deemed to be negligible for the demonstrated wavelength channel separation of 1.3 nm. The modulator array comprised four microring resonators configured as P-I-N carrier injection devices [61, 62]. The microring resonators were designed to have 4.98-, 5.00-, 5.02-, and 5.04-μm diameters, corresponding to an even wavelength channel spacing of 3.6 nm, with optical quality factors of about 20,000. The waveguides were 450-nm wide and 200-nm tall, with a 50-nm slab layer that was doped to form the P-I-N diode structures.

Another experimental demonstration showcased a silicon microring resonator electro-optic modulator array with four modulators, each encoded with 12.5-Gb/s data [60]. The modulators in this array were similar to the one presented in Sects. 3.4.1.1 and 3.4.1.2, with specified radius offsets of 20, 40, and 60 nm, corresponding to wavelength channel spacing of 3 ± 1 between the operating wavelengths of 1550.0 and 1561.2 nm.

The conclusions that are drawn from these experimental demonstrations are that these silicon microring resonator electro-optic modulator arrays:

- Are methodically scalable in the number of utilizable wavelength channels in the wavelength-parallel optical data stream, simply with the addition of more modulators in the array.
- Offer superior resilience to optical signal distortions caused by inter-channel crosstalk when scaling the number of utilized wavelength channels.
- Offer practical bandwidth scalability by increasing the modulation rate of each individual modulator. It is not unreasonable to envision similar modulator arrays with modulators each operating at 18 Gb/s [2], or more.

3.4.2.1 Inter-Modulation Crosstalk in Crystalline Silicon Microring Resonator Electro-Optic Modulator Arrays

The number of utilized wavelength channels in the photonic transmission system has a large dependence on the spectral density of these wavelength channels. Since each modulator in the modulator array only affects a single target wavelength channel, this alleviates the necessity of spectrally demultiplexing and multiplexing each wavelength channel for independent modulation. However, since the microring resonator modulator encodes the optical data by shifting its resonance in and out of the target wavelength channel, neighboring wavelength channels will suffer from inter-modulation crosstalk when the wavelength channel separation is too small.

Using the device presented in Sects. 3.4.1.1 and 3.4.1.2, the effects of inter-modulation crosstalk in the modulator array were experimentally characterized and quantified [63]. The results showed that there was no added power penalty due to inter-modulation crosstalk when the externally-modulated optical signal was placed as close as 50 GHz above, and 100 GHz below, the wavelength channel encoded using the microring resonator modulator [63]. This asymmetry corresponds to the blue shift of the resonance induced by the plasma-dispersion effect from injecting and extracting electrical carriers through the P-I-N diode integrated into the microring resonator of the modulator. When this externally-modulated optical signal was placed 25 GHz above the second wavelength channel, there was a corresponding power penalty caused by inter-modulation crosstalk. At 50 GHz below, this corresponding power penalty increased sharply.

To avoid signal distortion effects caused by inter-modulation crosstalk, the wavelength channel spacing, $\Delta\lambda$, corresponds to either the zero-power-penalty threshold below, $\Delta\lambda_{Below}$, or above, $\Delta\lambda_{Above}$, the measured wavelength channel, whichever value is greater. That is:

$$\text{If } \Delta\lambda_{Below} \geq \Delta\lambda_{Above}, \Delta\lambda = \Delta\lambda_{Below} \tag{3.1}$$

$$\text{If } \Delta\lambda_{Below} < \Delta\lambda_{Above}, \Delta\lambda = \Delta\lambda_{Above} \tag{3.2}$$

Combining the aforementioned experimental results with Eqs. 3.1 and 3.2, $\Delta\lambda_{Below} = 100\,\text{GHz}$ and $\Delta\lambda_{Below} = 50\,\text{GHz}$. Since $\Delta\lambda_{Below} < \Delta\lambda_{Above}$, $\Delta\lambda = 100\,\text{GHz}$, or $\Delta\lambda \approx 0.8\,\text{nm}$ at $\lambda = 1{,}550\,\text{nm}$, the operating wavelength. Further following Eqs. 3.1 and 3.2, the total utilized bandwidth in this wavelength-parallel system, $\Delta\lambda_{Total}$, given the maximum number of utilized wavelength channels, λ_{Max}, is described by:

$$\text{If } \Delta\lambda_{Below} \geq \Delta\lambda_{Above}, \Delta\lambda_{Total} = (\lambda_{Max} \times \Delta\lambda_{Below}) + \Delta\lambda_{Above} \tag{3.3}$$

$$\text{If } \Delta\lambda_{Below} < \Delta\lambda_{Above}, \Delta\lambda_{Total} = (\lambda_{Max} \times \Delta\lambda_{Above}) + \Delta\lambda_{Below} \tag{3.4}$$

The conclusions that are drawn here are that these silicon microring resonator electro-optic modulator arrays:

- Have a large dependence on inter-modulation crosstalk, arising from increasing the spectral density of the utilized wavelength channels, and subsequently reducing the wavelength channel spacing.
- Are capable of operating without any added optical signal distortion with 100-GHz wavelength channel spacing, which is about 0.8-nm at the 1,550-nm operating wavelength. Photonic transmission systems that are able to account for some signal distortion from inter-modulation crosstalk may further benefit from decreasing this wavelength channel spacing and increasing the number of utilized wavelength channels.

3.4.2.2 Bandwidth Scalability of Crystalline Silicon Microring Resonator Electro-Optic Modulator Arrays

These modulator arrays offer utilizable wavelength channel scalability that is determined and limited by two major factors: the FSR of the microring resonator, representing the maximum available bandwidth for the wavelength-parallel system, and the wavelength channel spacing utilized by the photonic transmission system. The FSR of a microring resonator is approximated by:

$$FSR \approx \frac{\lambda_0^2}{n_{Group} \times L} = \frac{\lambda_0^2}{n_{Group} \times 2\pi r} \tag{3.5}$$

Here, n_{Group} is the group index of the waveguide that forms the microring resonator, L is the optical path length for one round trip through the microring resonator, and r is the radius of the microring resonator. According to Eq. 3.5, in order to achieve the largest possible FSR, FSR_{Max}, the smallest possible optical path length, L_{Min}, and therefore the smallest possible microring resonator radius, r_{Min}, should be implemented, where:

$$FSR_{Max} \approx \frac{\lambda_0^2}{n_{Group} \times L_{Min}} = \frac{\lambda_0^2}{n_{Group} \times 2\pi r_{Min}} \tag{3.6}$$

Following Eqs. 3.3 and 3.4, the total utilized bandwidth in the wavelength-parallel system, $\Delta\lambda_{Total}$, must satisfy the following condition:

$$FSR_{Max} > \Delta\lambda_{Total} \tag{3.7}$$

Empirically, the smallest practical microring resonator electro-optic modulator that has been demonstrated, arranged in a five-modulator array configuration, had a 1.5-μm microring resonator radius with a measured optical quality factor of 9,000 [42]. The two main limitations associated with producing high-optical-quality-factor modulators with small microring resonator radii are the fundamental bending loss, caused by intrinsic modal leaking in the bent dielectric, and scattering loss, caused by sidewall roughness of the microring resonator waveguide. The achieved radius of 1.5 μm for the target optical quality factor is close to the theoretical limit of about 1.37 μm [42]. Furthermore, by experimentally measuring the FSR of the fabricated devices, which were about 62.5 nm, a group index of 4.2 was calculated.

From the aforementioned results, $n_{Group} = 4.2$ and $r_{Min} = 1.5\,\mu\text{m}$ ($L_{Min} \approx 9.4\,\mu\text{m}$) can be utilized for the wavelength $\lambda_0 = 1.55\,\mu\text{m}$. Using Eq. 3.6, given these parameters, $FSR_{Max} \approx 60.7\,\text{nm}$. The maximum number of utilized wavelength channels, λ_{Max}, may then be determined by dividing the maximum available bandwidth represented by the FSR, FSR_{Max}, by the wavelength channel spacing, $\Delta\lambda$:

$$\lambda_{Max} = \frac{FSR_{Max}}{\Delta\lambda} \tag{3.8}$$

The wavelength channel spacing then determines the number of available wavelength channels. In Sect. 3.4.2.1, it was shown that a practical wavelength channel spacing to avoid inter-modulation crosstalk is 100 GHz, which is about 0.8 nm at the 1,550-nm operating wavelength. Combining this with Eq. 3.8 produces $\lambda_{Max} = 75$ wavelength channels. If each wavelength channel is modulated at 18 Gb/s [2], this translates to a 1.35-Tb/s wavelength-parallel optical data stream. At 40 Gb/s per wavelength channel, as shown in Sects. 3.2.1.1 and 3.2.2.1, this corresponds to a 3-Tb/s stream. Furthermore, using Eqs. 3.3 and 3.4, $\Delta\lambda_{Total} \approx 60.5\,\text{nm}$, which is less than the maximum $FSR_{Max} \approx 60.7\,\text{nm}$, and satisfies the condition of Eq. 3.7.

3.5 Switches

Silicon photonic switches route high-bandwidth wavelength-parallel optical messages throughout the photonic interconnection networks. Since these networks present a dynamic switching environment, the switches used for routing must be able to not only handle large bandwidths, but also toggle between different states quickly and with low energy dissipation.

3.5.1 *Universal Microring Resonator Switches*

The microring resonator is a critical building block for compact and scalable silicon photonic systems. For a microring resonator coupled to two waveguides, as shown in Fig. 3.11, equations for the *through* port and *drop* port responses can be written down by inspection by considering the phase and amplitude of the optical mode as it propagates through the device [64].

Important parameters set by the material properties and waveguide cross-section are the propagation loss, α, and the effective index, $n_{Eff}(\lambda)$. The design parameters are the microring resonator radius, r, and the field transmission and coupling coefficients, t and κ, respectively. The coupling strength is determined by the separation between the waveguide and the microring resonator. A lossless coupler follows the equation:

$$t^2 + \kappa^2 = 1 \tag{3.9}$$

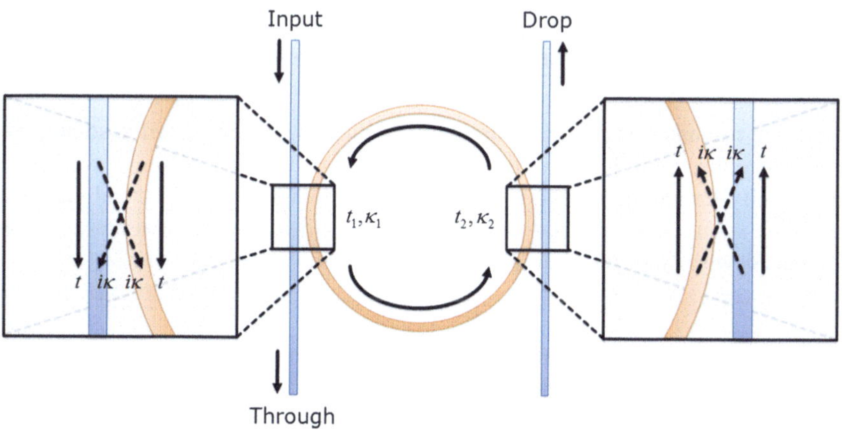

Fig. 3.11 Universal Description of the Microring Resonator Switch [24]: Schematic of a microring resonator coupled to *through* and *drop* waveguides, producing a microring resonator switch. *Close-up views* of the electric field coupling highlight the universal directional couplers. In a single pass, the transmitted component of the electric field is multiplied by t, and the coupled component is multiplied by $i\kappa$. This universal description is independent of the material system used

The quality factor, or spectral width, of the resonance is determined by the propagation loss and coupling. Similarly, the extinction ratio of the resonance is determined by the ratio of propagation loss to coupling, with critical coupling occurring for:

$$t_1 = t_2 \times \alpha \tag{3.10}$$

Tuning the filter wavelength (or modulating the optical transmission) is accomplished by changing the material refractive index, n, of the microring resonator, which changes the effective index, n_{Eff}, of the optical mode, which shifts the resonant wavelength. The relationship between material and effective index is given by:

$$\Delta n_{Eff} = \Gamma \times \Delta n \tag{3.11}$$

where Γ is the optical mode's silicon confinement factor [65], determined by the cross-sectional dimensions of the silicon waveguide, given by:

$$\Gamma = \frac{\partial n_{Eff}}{\partial n} \tag{3.12}$$

Tuning the refractive index may be accomplished on microsecond time scales by changing the temperature, or on sub-nanosecond time scales by changing the electron and hole concentrations using FCD [1], which may be implemented using an integrated electrical diode. Ultrafast tuning of the refractive index using FCD is generally also accompanied by insertion loss from FCA, which broadens the resonance and changes the critical coupling condition for the maximum extinction ratio.

3.5.2 Microring Resonator Broadband Switches

Microring resonator electro-optic broadband switches are considered in some photonic interconnection networks for their ultrahigh bandwidth and low energy dissipation. By leveraging comb-switching with wavelength-parallel optical message encoding, several experimental verifications have shown the potential these silicon photonic devices have for routing ultrahigh bandwidth signals.

3.5.2.1 Elementary Microring Resonator Broadband Switches

Combining the functional ubiquity of the microring resonator with the advantages of higher-order configurations [3, 66–69], broadband switching devices are realized with massive switching bandwidths and short switching transitions. In Fig. 3.12, the critical elementary models are defined for 1 × 2 [3, 66–70], and 2 × 2 [71], switching building blocks, implemented using arbitrary-order resonators, noting that different configurations are required for odd and even number of resonators. These elementary

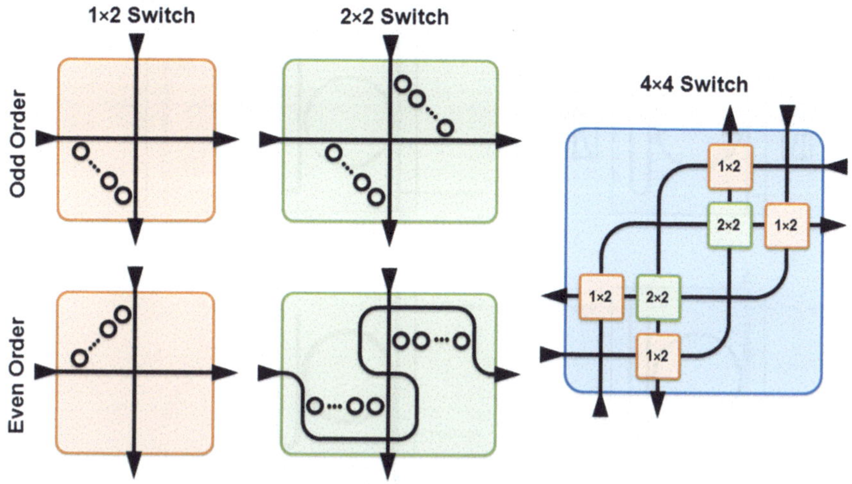

Fig. 3.12 Microring Resonator Broadband Switch Elementary Models [69]: Elementary models for 1×2 and 2×2 switching building blocks, implemented with both odd- and even-order microring resonators, as well as the non-blocking 4×4 switch constructed using these models

models are combined to form more complex switching subsystems, such as the non-blocking 4×4 switch, also shown in Fig. 3.12, a critical building block in many photonic interconnection networks [72].

3.5.2.2 All-Optical Switching

A broadband switch device comprising a microring resonator with resonant modes spaced by approximately 100 GHz has been demonstrated and characterized. This spacing allows the microring resonator to operate as a comb switch on a broadband, wavelength-parallel data stream in much the same way a smaller-diameter microring resonator affects a single-channel optical signal. This type of switch, which was shown to easily support aggregate data rates of 160 Gb/s, is intended for use in high-performance photonic interconnection networks [73].

One such device was a broadband all-optical comb switch, comprising a microring resonator coupled to two straight waveguides; one is the input port and *through* port, and the other is the *drop* port, as depicted in Fig. 3.13. On resonance, light is coupled into the microring resonator, and is sent to the *drop* port. Off resonance, light propagates virtually unaffected to the *through* port. Leveraging the relatively small FSR, multiple channels can be switched simultaneously for broadband photonic interconnection network routing applications.

The high performance of the microring resonator, in terms of resonant mode uniformity and wavelength crosstalk, indicates that significantly more wavelength channels could be switched using this device. Thus, the aggregate signal bandwidth that

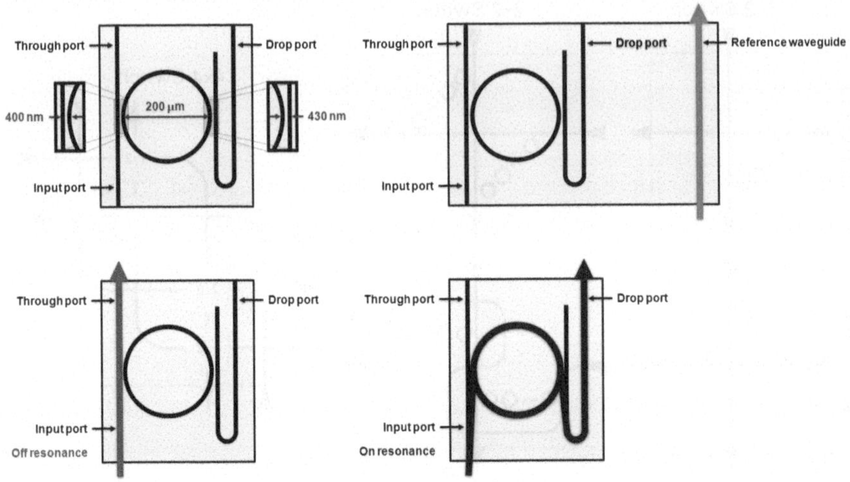

Fig. 3.13 1 × 2 Microring Resonator Broadband Switch Schematic: Schematic representations of optical signal paths through the reference waveguide and for two possible states of the microring resonator broadband switch, along with the typical dimensions of the device

can be routed through the device depends on the allowable data rate per wavelength channel. Devices designed with wider bandwidths would enable data rates of 40 Gb/s per wavelength channel or more, scaling the bandwidth well into terabit-per-second data rates.

The conclusions that are drawn from these experimental demonstrations are that these 1 × 2 silicon microring resonator broadband switches:

- Have the ability to propagate error-free optical data streams, with an aggregate bandwidth of at least 160 Gb/s.
- Show no increase in power penalty from wavelength crosstalk for up to 16 wavelength channels, a property that will clearly enable more wavelength channels to be transmitted through the device.
- Exhibit achievable bandwidth near one terabit per second, but even higher aggregate bandwidths are envisioned for future devices.

Wavelength-parallel all-optical switching was demonstrated using this device [74]. When no optical pump was applied, input light on resonance with the microring resonator was coupled to the *drop* port of the device, and light that was off resonance propagated to the *through* port. The wavelengths of the microring resonator's resonant modes may all be blue-shifted simultaneously by injecting electrical carriers into the device through the free-carrier plasma dispersion effect. As a result, when the wavelength of an optical data signal is aligned on resonance, the presence of a carrier-generating pump source switches the signal from the *drop* port to the *through* port. Likewise, the removal of the carriers directs the signal back to the *drop* port.

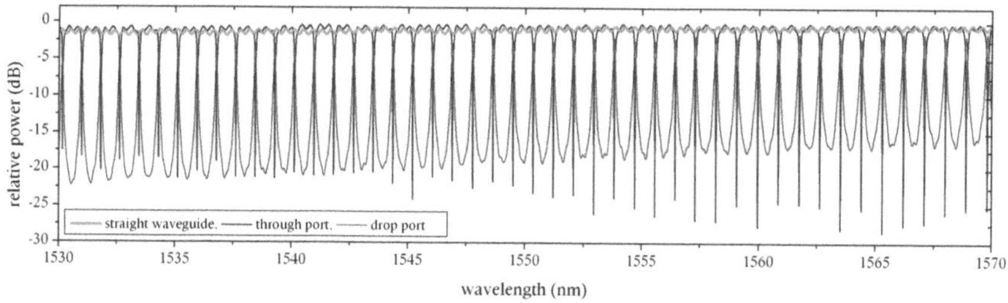

Fig. 3.14 1 × 2 Microring Resonator Broadband All-Optical Switch Spectra with 47 Resonant Modes [74]: Relative transmission spectra of the *through* and *drop* ports of the broadband switch, and the output port of the reference waveguide, which is used to evaluate the insertion losses of the broadband switch

Carriers may be injected using an optical pump, or an electrical signal applied across a P-I-N junction surrounding the microring resonator waveguide.

All-optical switching of two CW wavelength channels has been previously demonstrated [75]. A 160-Gb/s optical data stream, comprising 16 wavelength channels each modulated at 10 Gb/s, was passed through the switch passively (with no applied pump), and the power penalty due to inter-channel crosstalk within the microring resonator was found to be negligible when scaling from one to 16 wavelength channels [73]. In subsequent work, full switching of 20 wavelength channels simultaneously was demonstrated [74].

The transmission spectra of both output ports of the switch, along with a third port (a reference waveguide which has the same length and cross-section as the *through*-port waveguide), as shown in Fig. 3.13, were plotted relative to each another in Fig. 3.14. The measured power from the *through* port was comparable to that of the reference waveguide, indicating negligible *through* port insertion loss beyond the waveguide propagation loss.

The switching ratio and the transition times of the probe after exiting the *drop* port, a shown in Fig. 3.15, may be improved with a more optimal pump configuration, and the transition times can be further reduced to less than 100 ps using electro-optic switching.

The conclusions that are drawn from these experimental demonstrations are that these 1 × 2 silicon microring resonator broadband switches also:

- Are capable of simultaneously switching at least 20 CW wavelength channels, spanning 25 nm (Fig. 3.16).
- Exhibit error-free active operation at near-gigahertz speeds, for single-channel optical data at 10 Gb/s. This device should also have the capability to switch wavelength-parallel optical data with comparable performance.

Furthermore, all-optical simultaneous high-speed switching of a 250-Gb/s optical signal, using 20 wavelength channels, each modulated at 12.5 Gb/s was demonstrated [76]. The resulting switching extinction ratio was examined with varying pump powers. As shown in Fig. 3.17, the wavelength channels, spanning more than 25 nm,

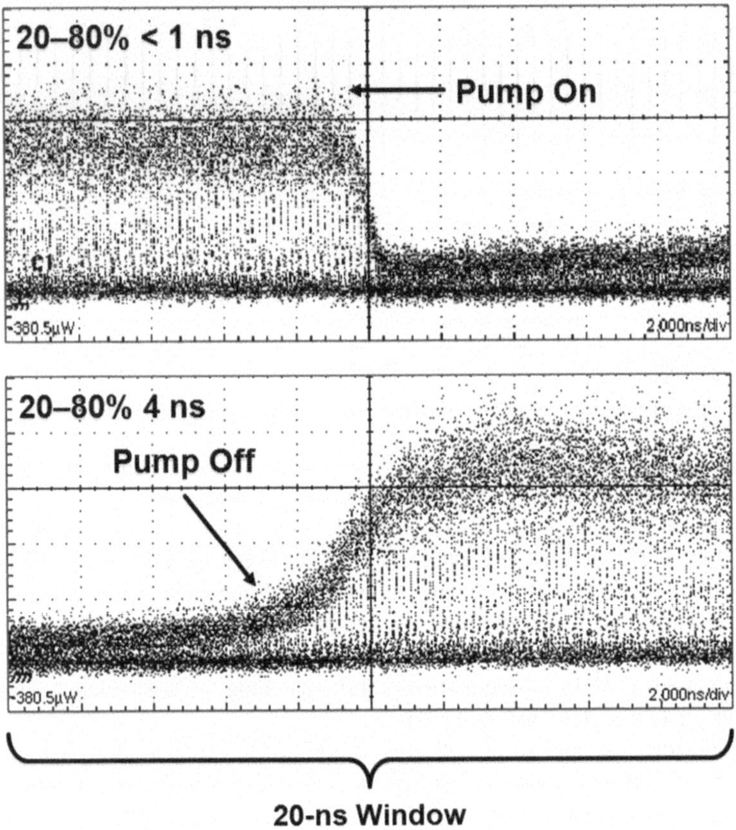

Fig. 3.15 Output Optical Data Packets During Single-Channel Switching of 1 × 2 Microring Resonator Broadband All-Optical Switch [74]: Output optical packets encoded with 10-Gb/s data egressing from the *drop* port of the broadband switch, illustrating the extinction ratios and switching times

comprised channels C21–C27, C33–C38, and C47–C53 of the *ITU* grid. Similar reported devices provide a promising path toward alleviating these tight wavelength requirements by providing broader per-channel bandwidths [68].

Extinction ratios were measured at both ports for varying injected pump powers, as shown in Fig. 3.18. The extinction ratios were improved on both ports with larger pump powers. At the *drop* port, the extinction ratio was improved by 6.5 dB with the added 12 dBm of pump power; at the *through* port, the extinction ratio was improved by 11.5 dB. The switching ratios and transition times, will be further improved with a more optimal pump configuration using an electrical signal applied across a P-I-N junction surrounding the microring resonator waveguide.

Here, the main conclusions that are drawn are that these 1 × 2 silicon microring resonator broadband switches also:

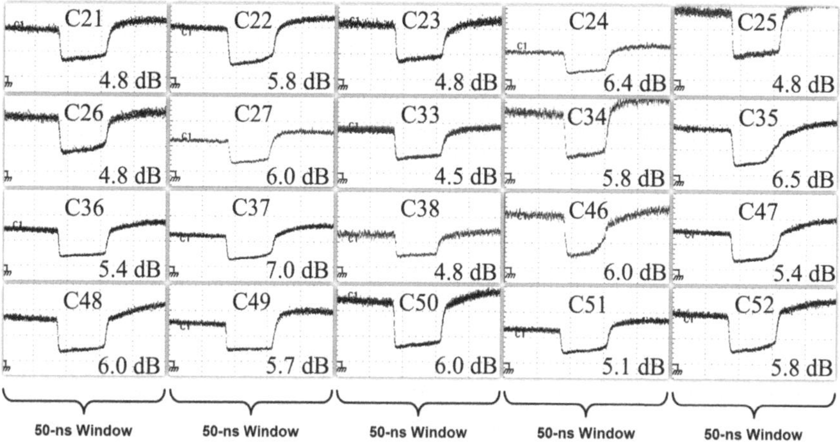

Fig. 3.16 Wavelength-Parallel Switching of 1 × 2 Microring Resonator Broadband All-Optical Switch with 20 Wavelength Channels [74]: Temporal waveforms egressing from the *drop* port of the broadband switch under active operation, with corresponding labels of the *ITU* grid wavelength channels and extinction ratios

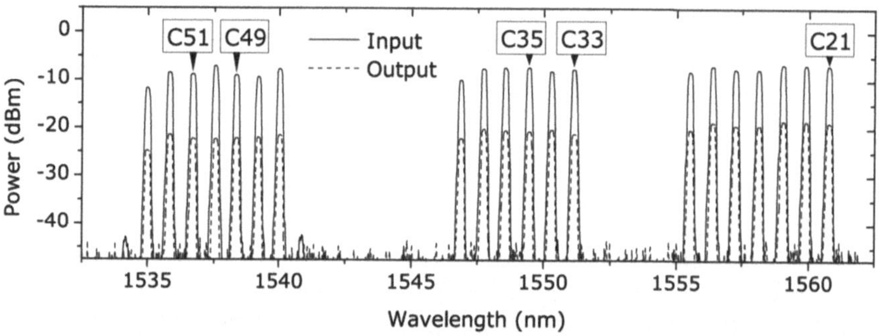

Fig. 3.17 1 × 2 Microring Resonator Broadband All-Optical Switch Spectra with 20 Utilized Resonant Modes [76]: 250-Gb/s transmission spectra utilizing 20 wavelength channels, each encoded with 12.5-Gb/s optical data, before being injected into the chip, and at the *drop* port of the broadband switch

- Are indeed capable of error-free all-optical simultaneous switching of wavelength-parallel optical signals, consisting of at least 250-Gb/s aggregate data, using 20 wavelength channels that span more than 25 nm.
- Are capable of being optimized for performance by increasing the pump power to maximize the switching extinction ratio. Using a direct electrical signal to generate the free carriers at the microring resonator will further improve the extinction ratios, switching times, energy dissipation, and overall scalability, of this broadband switch.

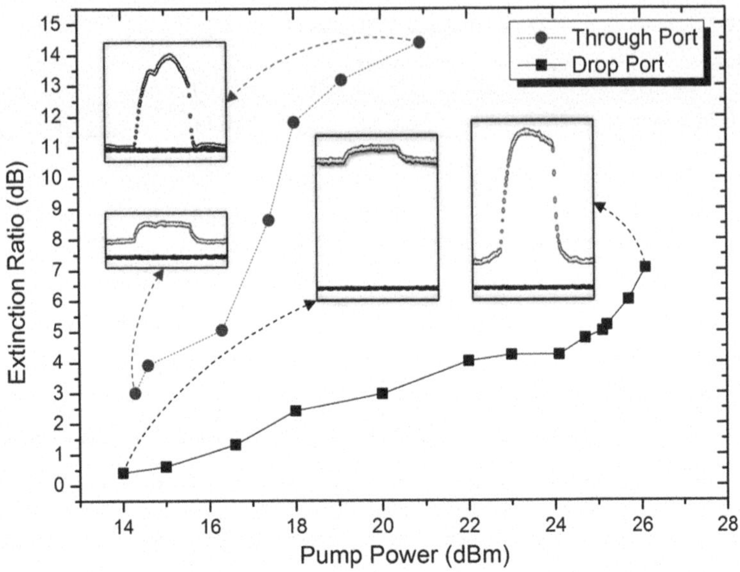

Fig. 3.18 Extinction Ratios During All-Optical Switching of 1 × 2 Microring Resonator Broadband All-Optical Switch [76]: Experimentally-measured extinction ratios at both output ports during the switching of the broadband switch, for varying optical pump powers

3.5.2.3 Electro-Optic Switching

Within the photonic interconnection networks, broadband electro-optic switches route high-bandwidth wavelength-parallel optical messages throughout the dynamic circuit-switched paths with ultrafast reconfiguration times [31]. Combining the versatility of the silicon photonic resonator with the advantages of higher-order configurations [3, 66–69], it is possible to realize broadband switching devices with massive switching bandwidths and short switching transitions. The critical elementary models were first defined for 1 × 2 [3, 66–70, 73–76], and 2 × 2 [71, 77], switching building blocks, implemented using arbitrary-order resonators, noting that different configurations are required for odd [68, 70–82], and even [3, 66, 67, 69], number of resonators, shown in Fig. 3.12.

These elementary models can be combined to form more complex switching subsystems, such as the non-blocking 4 × 4 switch, a critical building block in many photonic interconnection networks [72, 78–82]. A silicon photonic microring resonator electro-optic switch was demonstrated experimentally dynamically routing single-channel data rates up to 40 Gb/s, showcasing high bandwidth, short switching transitions, high extinction ratios, and low driving voltage. The device was a second-order 1 × 2 switch, consisting of two coupled microring resonators each coupled to a waveguide, as shown in Fig. 3.19.

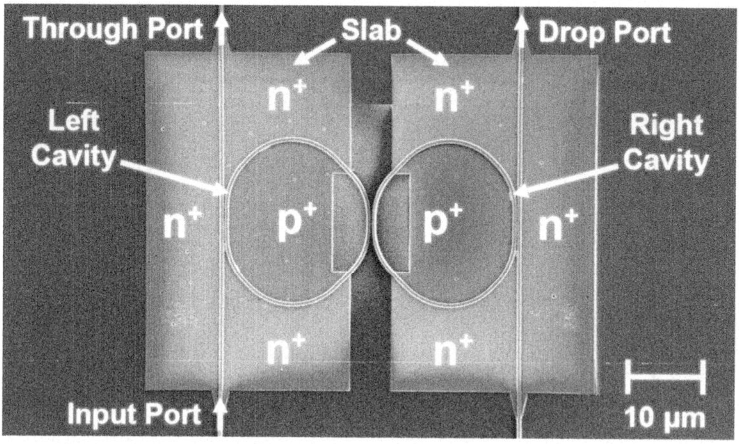

Fig. 3.19 Scanning-Electron-Microscope Image of 1 × 2 Microring Resonator Broadband Electro-Optic Switch [67]: *Top-view* scanning-electron-microscope (SEM) image of the silicon photonic microring resonator electro-optic switch

As seen in Fig. 3.19, the microring resonators in the 1 × 2 switch were designed with both racetrack and ring features, with $2\pi \times 10$-μm cavity lengths [3]. The waveguides were 450-nm wide and 250-nm tall; there was a 40-nm slab near the microrings that was doped to form the P-I-N diode structures. Switching an optical signal between the *through* port and the *drop* port was accomplished with the detuning of the right cavity resonance using the free-carrier dispersion effect arising from injecting and extracting electrical carriers through the P-I-N diode. Electro-optic control of these switches enables a more scalable and energy-efficient interconnection network compared to the all-optical switching methods demonstrated in previous work [68, 70, 71, 73–77].

Before electrically driving the switch, the spectrum of this device was recorded for both output ports in the passive state, as depicted in Fig. 3.20, observing a 9-nm FSR and *through* port passbands with 70-GHz 3-dB bandwidths. The passbands of the two cavities were not perfectly overlapping in this passive state; these passbands were aligned with the applied voltage bias during active switching, and have been shown to achieve depths greater than 20 dB [3]. A high-speed data signal was then injected at the input port of the switch with 5-, 10-, 20-, and 40-Gb/s data rates, with recorded eye diagrams of the optical signal egressing from the *through* port (at 1561.5 nm) and *drop* port (at 1559.5 nm) of the switch, shown in Fig. 3.21. These eye diagrams are compared with the back-to-back case, which bypassed the silicon chip.

For active switching, the optical signal was first aligned to be on resonance at 1559.5 nm. When the voltage signal was set *high* (*low*), the optical signal was switched to the *through* port and *drop* port. The device was actively switched with a 1.3-V_{PP} square wave with a 0.5-V voltage bias, and a 100-ns period with a 50 % duty cycle, producing 50-ns optical data packets alternately egressing from each output

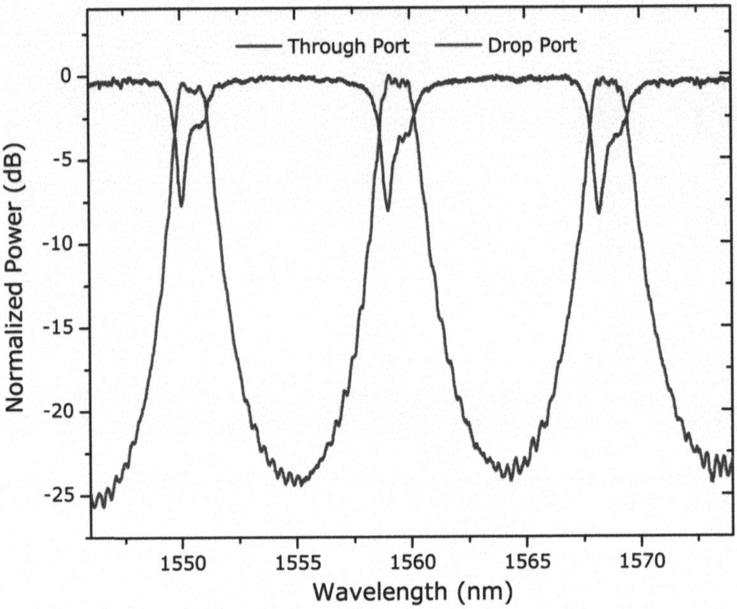

Fig. 3.20 Spectra of Resonant Response of 1 × 2 Microring Resonator Broadband Electro-Optic Switch [67]: Spectra of the resonant response for both output ports of the silicon photonic microring resonator electro-optic switch

port. An optical signal encoded with 40-Gb/s data was switched, and optical data packets were recorded, including their rising and falling edges, at each output port of the switch, shown in Fig. 3.22, observing greater than 12-dB extinction ratios at both output ports. The sub-nanosecond rising edges are achieved using the P-I-N diode structure. The falling edges are typically limited by carrier lifetimes, and can be further improved using the pre-emphasis method [83].

Two microring resonators can be combined to build a 2 × 2 broadband switch, an important building block of a high-performance photonic interconnection network [71, 77]. The device used for these experiments was fabricated using EBL and RIE on top of a 3-μm-thick buried oxide layer of an SOI substrate, and covering the resulting structure with a 3-μm-thick silicon dioxide overcladding layer using plasma-enhanced chemical vapor deposition (PECVD). The microring resonators in this 2 × 2 switch had 100-μm diameters, corresponding to an FSR of 1.6 nm. The waveguides were 450-nm wide and 250-nm tall, and utilized inverse-tapers at the chip facets. In each experiment, the resonant modes of the microring resonators, which have 3-dB bandwidths of 0.1 nm (12.5 GHz) to accommodate 10-Gb/s optical signals, are aligned to spectrally overlap with local thermo-optic tuning.

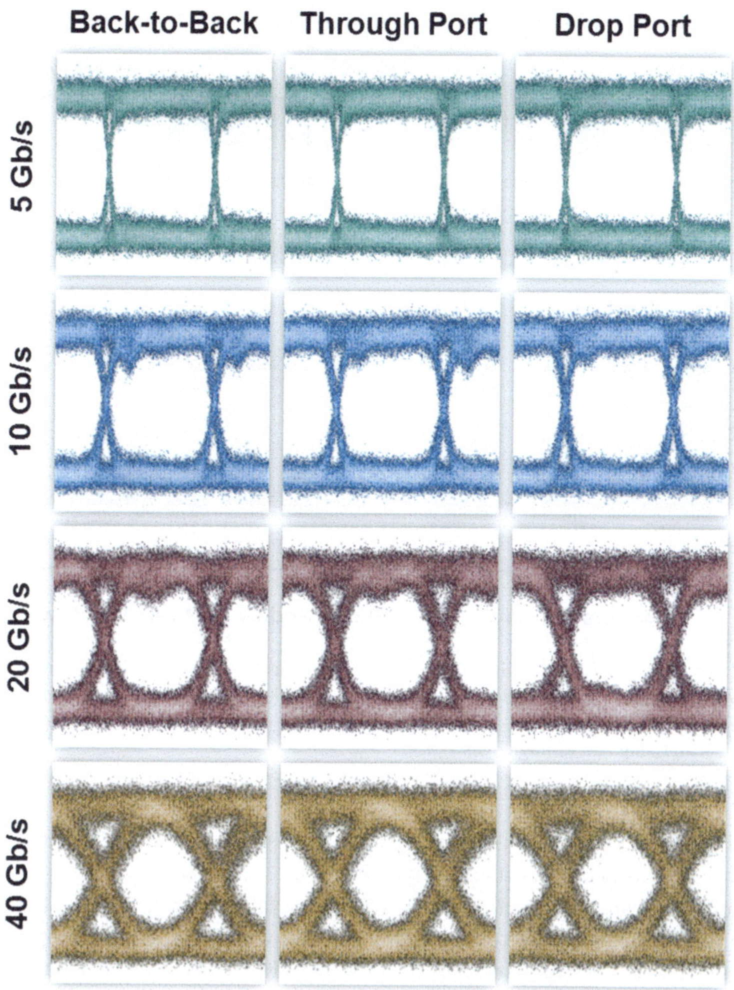

Fig. 3.21 Eye Diagrams of Switching of 1 × 2 Microring Resonator Broadband Electro-Optic Switch [67]: Output eye diagrams for optical signals with 5-, 10-, 20-, and 40-Gb/s data rates, egressing from both output ports of the silicon photonic microring resonator electro-optic switch, as well bypassing the silicon chip in the back-to-back case

3.5.2.4 Static Measurements for Wavelength-Parallel All-Optical Switching Using 2 x 2 Silicon Microring Resonator Broadband Switch

A static experimental demonstration of this broadband switch was accomplished for a wavelength-parallel optical signal [77]. This comprised six wavelength channels each encoded with 10-Gb/s optical data and tuned to have the average optical power of −6 dBm. Power penalties were measured for all six wavelength channels for the

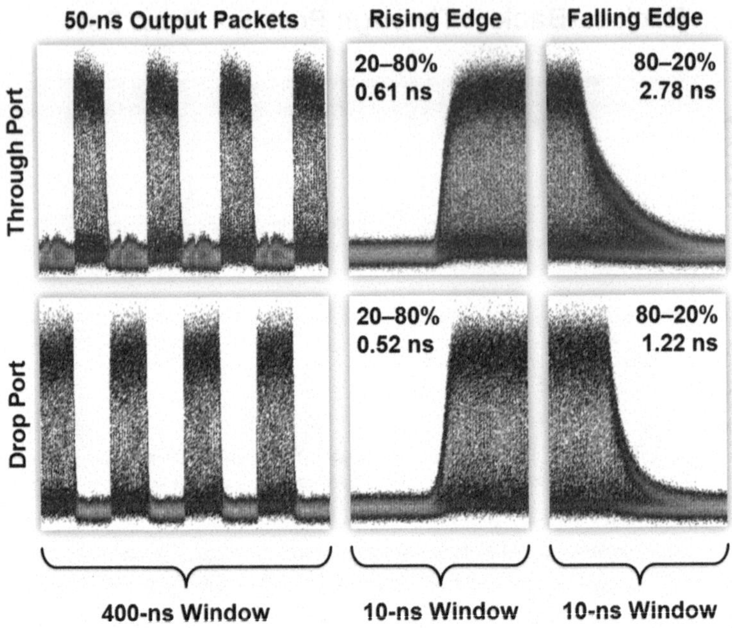

Fig. 3.22 Optical Packets from Switching of 1 × 2 Microring Resonator Broadband Electro-Optic Switch [67]: Output optical packets encoded with 40-Gb/s optical data for both output at the *through* and *drop* ports of the silicon photonic microring resonator electro-optic switch, while switching, with rising and falling edges

bar and *cross* states of the broadband switch, ranging from 0.1 to 1.2 dB and 0.0 to 0.5 dB for the *bar* and *cross* states, respectively.

The effects of inter-channel crosstalk were then evaluated by measuring the power penalty for varying average optical powers of the wavelength-parallel signal, and comparing the effects with those observed for a single channel with the same varying average optical powers [77]. The average optical powers that were measured were 6, 11, 15, 18, and 23 mW, all producing negligible effects on the power penalty when switching between the single-channel and wavelength-parallel configurations.

The main conclusions that are drawn from these experiments are that these 2 × 2 silicon microring resonator broadband switches:

- Are capable of error-free transmission of wavelength-parallel optical signals, consisting of at least 60-Gb/s aggregate data, using six wavelength channels, each encoded with 10-Gb/s optical data.
- Show negligible inter-channel crosstalk for at least six wavelength channels, for total average optical powers of at least up to 23 mW.

3.5.2.5 Dynamic Measurements for Wavelength-Parallel All-Optical Switching Using 2 × 2 Silicon Microring Resonator Broadband Switch

Optical signals traversing the device are first aligned to the resonances of the microring resonators. An applied optical pump toggles the state of the switch by simultaneously tuning all the resonances away from the optical signal wavelengths, changing the switch configuration from the *bar* to the *cross* state [71, 77]. Switching this optical pump off changes the switch configuration from the *cross* to the *bar* state. This optical pump comprises two lightwaves, co- and counter-propagating, with average powers of 18 and 17 dBm, respectively. TPA is used here to induce the electrical carriers required to shift the resonances through FCD.

Dynamic experimental demonstration with this broadband switch has yielded switching transitions below 2 ns, and extinction ratios as high as 11.5 dB [71, 77]. This was accomplished for a wavelength-parallel optical signal comprising six wavelength channels each encoded with 10-Gb/s optical data. The crosstalk between the two output ports was measured to be between 12.3 and 7.0 dB for the *bar* and *cross* states, respectively. Furthermore, the power penalty was measured to be 1.9 and 3.5 dB for the *bar* and *cross* states, respectively.

In this work, the main conclusions that are drawn are that these 2 × 2 silicon microring resonator broadband switches also:

- Are capable of error-free all-optical switching of wavelength-parallel optical signals, consisting of at least 60-Gb/s aggregate data, using six wavelength channels, each encoded with 10-Gb/s optical data.
- Exhibit 2-ns switching times and 11.5-dB extinction ratios during the all-optical switching of wavelength-parallel optical signals, which can be further improved with direct electrical carrier injection using P-I-N diode structures.

3.5.3 Microring Resonator 4 × 4 Broadband Switches

1 × 2 and 2 × 2 switches, waveguides, and waveguide crossings can be combined to form a 4 × 4 photonic switch, which directs an input from any of the four ports to any of the other three available output ports without obstructing the paths of signals incident from other ports [82]. Using this switch, rather than a simpler blocking switch, improves performance and simplifies the network-level routing algorithms necessary to minimize (or entirely eliminate) packet contentions. 4 × 4 photonic switches are extensively described in Sect. 5.2.2.

The 4 × 4 switch, shown in Fig. 3.23 was fabricated on an SOI wafer, using EBL and RIE. All waveguides, including those constructing the microring resonators, are 450-nm wide and 250-nm tall. The gaps between the microring resonators and straight waveguides are 200 nm throughout, waveguide crossings are adiabatically tapered to 2-μm widths at the intersection to minimize reflections, and the microring resonators have 20-μm diameters, corresponding to 8-nm FSRs.

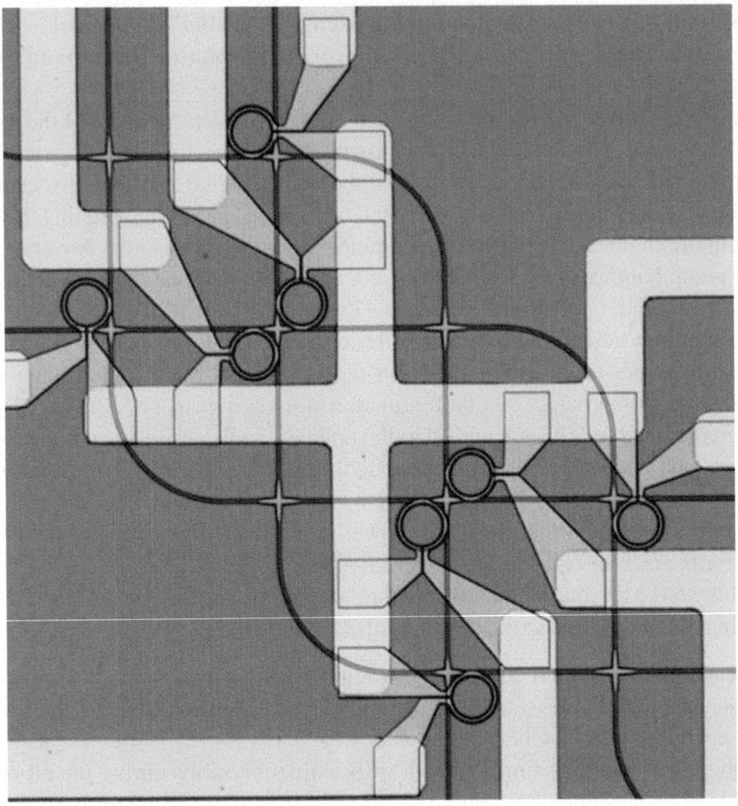

Fig. 3.23 Microscope Image of 4×4 Microring Resonator Broadband Router [82]: Microscope image of the fabricated non-blocking four-port bidirectional multi-wavelength message router

The structure provides a dedicated path for every I/O combination except U-turns. The router employs eight microring resonators, each coupled to a waveguide crossing, comprising four 1×2 switches [70], with a single microring resonator at a crossing, and two 2×2 switches [71], with two microring resonators at a crossing. The electronic heaters provide static switch-state configuration using DC resonance tuning. The four router I/O ports are referred as *North* (N), *East* (E), *South* (S), and *West* (W), corresponding to their relative positions.

The switch utilized metal heaters, fabricated about 1 μm above each microring resonator, on top of the oxide over-cladding layer. The metal traces are 1-μm wide and 300-nm tall, and are each used to tune the wavelengths of the resonant modes in order to mitigate fabrication imperfections.

When no thermal tuning was performed on the microring resonators, the optical signal bypassed each of the four microring resonators, egressing from port W. Thermally tuning the necessary microring resonator aligns its resonant mode to the signal wavelength.

The conclusions that are drawn from these experimental demonstrations are that these 4×4 silicon microring resonator non-blocking broadband switches:

- Are indeed feasible to implement silicon photonic routers in photonic interconnection networks.
- Are only required to switch their state at the packet rate, instead of the data rate. This has the potential to amortize the dynamic energy dissipated in state transients.
- May utilize the energy-efficient thermal tuning to mitigate fabrication imperfections. Electro-optic control of the microring resonators in these switches will enable the demonstrated functionalities with sub-nanosecond switching speeds.
- Have the capacity to support single-channel optical signals with data rates of at least 12.5 Gb/s. Wavelength-parallel operation of this switch will enable the routing of much larger aggregate bandwidths, with no increase in the required switching energy.

One of the most prevalent photonic element in 2D network topologies is the non-blocking four-port bidirectional router, which dynamically routes broadband messages to their target destinations. These networks utilize wavelength-parallel message encoding for optical-domain bandwidth enhancement. That is, multiple wavelength channels are passed through the network from source to destination cohesively, as a combined data-carrying unit, while other messages in the network circumvent contention using spatial avoidance. The network performance in every topology, as expected, has been shown to be vastly improved when multiple multi-wavelength messages can simultaneously pass through the photonic routers without contention. Therefore, the routing elements in these topologies must ideally provide four bidirectional I/O ports, with multi-wavelength routing capabilities and strictly non-blocking functionality.

Such a router has been designed, and its network-level performance has been evaluated through simulations [79, 84]. Furthermore, the router was fabricated [80], and initial data characterizations were performed [82]. This device was then fully characterized as a 30-Gb/s wavelength-parallel silicon photonic router, comprising three wavelength channels each modulated at 10 Gb/s [72, 81].

The waveguides are equipped with inverse-taper mode converters at each chip edge, and at one side of the chip a tapered-lensed fiber was employed to gather light from both the *East* and *South* output ports, alternately. At the other side of the chip, a lens was used to collimate light exiting from the *West* port while simultaneously focusing light entering into the *North* port. A polarizer transmitted the transverse electric (TE) polarization component of the two collimated beams.

Both single- and multi-wavelength experiments were performed to demonstrate the operation of the router. Here, the signal that egressed from the *East*, *South*, and *West* output ports was monitored while injecting into the *North* input port, as shown in Fig. 3.24. First, pattern traces of the single-channel signal exiting the destination output ports, along with the noise observed on the other output ports, are illustrated in Fig. 3.25, indicating extinction ratios in excess of 8 dB for each port.

Utilizing three consecutive resonance modes of the microring resonators, which have FSRs of approximately 8 nm and 3-dB bandwidths of about 38.5 GHz,

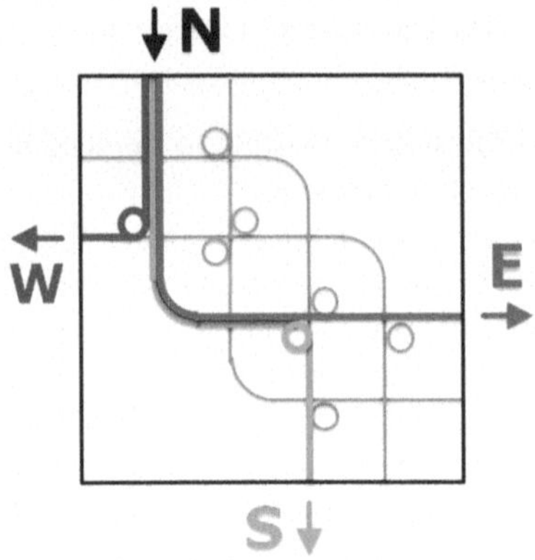

Fig. 3.24 Experimental Configurations of Wavelength-Parallel Switching Using 4 × 4 Microring Resonator Broadband Router [81]: Schematic of the broadband router, highlighting the signal paths of the three experimental switch configurations: $N \rightarrow E$, $N \rightarrow S$, and $N \rightarrow W$

a three-channel wavelength-parallel signal was then routed through the same switch configurations described above. The eye diagrams (Fig. 3.26) for the three-channel signal injected into the *North* input port, as well as at each destination output port for each experimental switch configuration, were recorded for each of the three 10-Gb/s

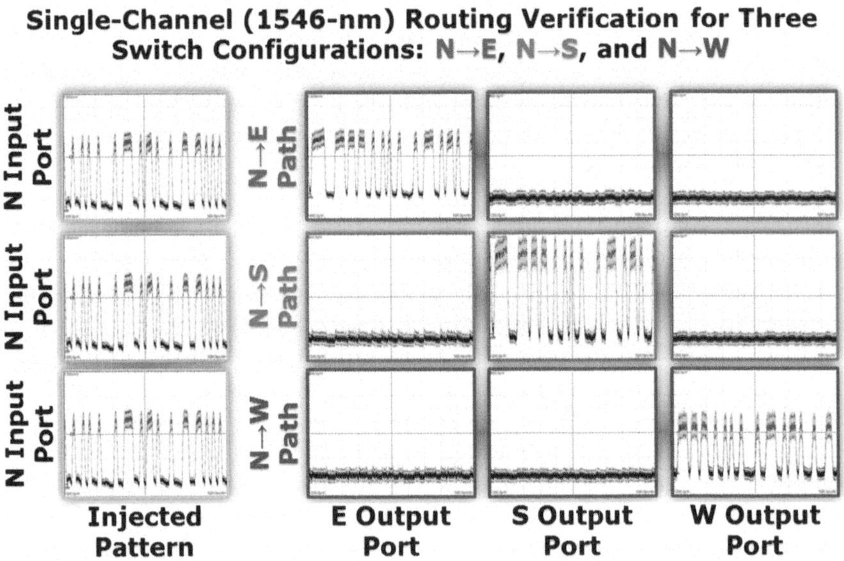

Fig. 3.25 Single-Channel Verification of Routing Functionality Using 4 × 4 Microring Resonator Broadband Router [81]: Pattern traces depicting single-channel (1546-nm) verification of the routing functionality for three experimental broadband router configurations: $N \rightarrow E$, $N \rightarrow S$, and $N \rightarrow W$. In each configuration, a single-wavelength signal (injected pattern) is injected into the *North* (N) input port, and examined on each output port (E, S, and W). Each trace spans 5 ns and has an amplitude scale of 100 μW/division, except for the input pattern and the *East* (E) output pattern under the $N \rightarrow E$ configuration, which are 200-μW/division

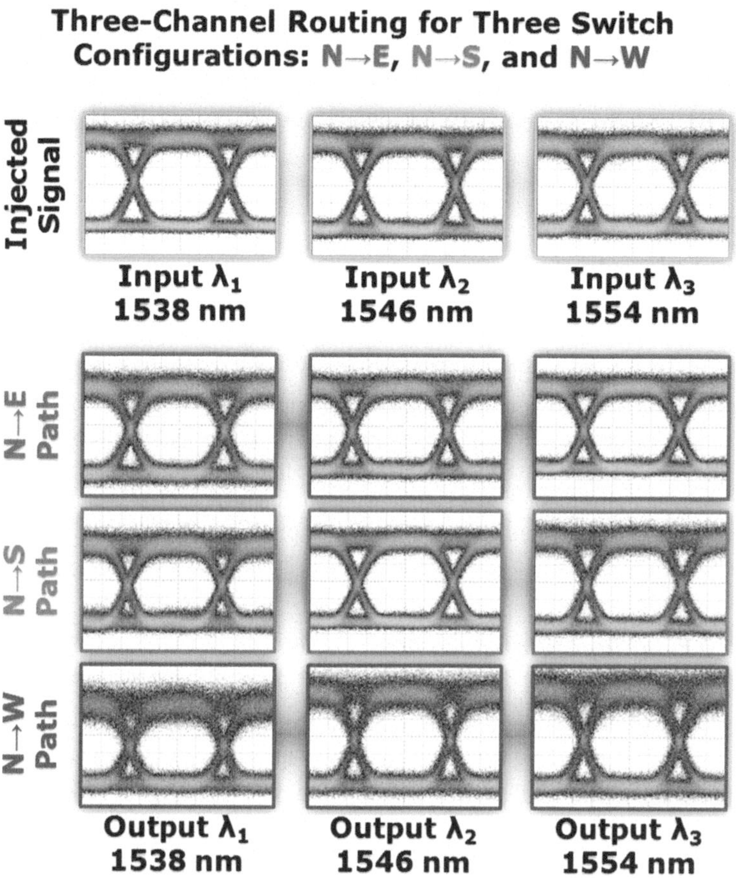

Fig. 3.26 Eye Diagrams of Wavelength-Parallel Routing Using 4 × 4 Microring Resonator Broadband Router [81]: 10-Gb/s eye diagrams for the wavelength parallel input signal (injected signal) injected into the *North* (N) input port, with eye diagrams for the wavelength-parallel output signal at the destination output port, for each experimental switch configuration: $N \rightarrow E$, $N \rightarrow S$, and $N \rightarrow W$. The wavelength-parallel signal consists of three 10-Gb/s wavelength channels, located at 1538, 1546, and 1554 nm. Each eye diagram has a 200-ps time span

wavelength channels, which were located at 1538, 1546, and 1554 nm. The power penalties, defined as the degradation in receiver sensitivity incurred by the device under test, were approximately 1.3 dB for the $N \rightarrow W$ state, at the BER of 10^{-9}. Both of the other states demonstrated power penalties below 1 dB for all three wavelength channels.

Here, the conclusions that are drawn from these experimental demonstrations are that these 4 × 4 silicon microring resonator non-blocking broadband switches also:

- Have the capacity to support wavelength-parallel optical signals. Successful transmission of 30-Gb/s aggregate data rates, consisting of three wavelength channels

each operating at 10 Gb/s has already been demonstrated [72]. Increasing the wavelength parallelism and single-channel data rates will further increase the aggregate data rate that is routed using this switch to the terabit-per-second scale.
- Manifest error-free performance for routing through each possible configuration of the switch.

3.6 Photodetectors

Located at the end of the optical communication link, the photodetector translates an incoming high-speed optical signal into the electrical domain. This is accomplished by absorbing the light to generate an electrical current. In many photonic interconnection network designs, crystalline silicon microring resonator filters are used to demultiplex wavelength-parallel optical data into spatially-parallel optical data, placing each wavelength channel on a unique waveguide. Once each wavelength channel is isolated, it can be detected using a high-speed photodetector.

3.6.1 Germanium Photodetectors

Since crystalline silicon does not exhibit linear absorption in the telecommunication wavelength bands, photonic interconnection networks have leveraged germanium as the absorbing material for photodetection [19]. Recent efforts in integrating germanium photodetectors with crystalline silicon waveguides have yielded many high-performance CMOS-compatible devices targeting high bandwidth, high responsivity, high quantum efficiency, low dark current, and low capacitance. Leveraging this material system, bandwidths exceeding 40 GHz [85–87], responsivities above 1 A/W [85, 88], quantum efficiencies higher than 90 % [87, 88], dark currents below 200 nA [88], and capacitances around 2 fF [87], have all been demonstrated.

3.6.2 Silicon Photodetectors

Another emerging method for producing photodetectors is to use silicon with crystal defects as the absorbing material. Recent efforts have enhanced linear absorption by implanting ions to create defects [89], obtaining bandwidths exceeding 35 GHz and responsivities as high as 10 A/W in different devices [90]. Similarly, deposited polycrystalline silicon can be used as the absorbing material in a microring resonator geometry to enhance absorption and reduce the footprint, demonstrating responsivities as high as 0.15 A/W [14]. Due to the moderate optical absorption, the absorbing polycrystalline silicon material can serve as both the demultiplexing filter and the photodetector.

3.6.3 Photodetector Arrays

Several photodetectors, each tuned to a distinct wavelength channel, can be linked to yield a photodetector array. This technique is similar to what has recently been demonstrated with four cascaded crystalline silicon microring resonator filters demultiplexing a wavelength-parallel optical signal into germanium photodetectors with 19-GHz bandwidths [87].

3.7 Lasers

All photonic transmission systems employ lasers for generating light at specific target wavelengths, which is then encoded with data and transmitted through the photonic network. Lasers emit light through a process of optical amplification based on the stimulated emission of photons. Most lasers consist of a gain medium, a pump, and a mechanism for optical feedback. The gain medium is a material with properties that allow it to amplify light at specific wavelengths through stimulated emission. The pump supplies the required energy to the gain medium, typically in the form of electrical current or light at a different wavelength. Optical feedback is commonly achieved using an optical cavity. An optical cavity is a structure that causes light to bounce back and forth, such as the space between two parallel mirrors. By placing the gain medium inside the optical cavity, light will repeatedly pass through the gain medium and be amplified with each pass.

For photonic interconnection networks, key laser parameters such as wavelength of operation, maximum output power, power efficiency, stability, footprint, CMOS compatibility, and cost determine the feasibility of each technology. One key metric for a laser technology is its wall-plug efficiency, which may render an otherwise feasible laser technology prohibitive in the system.

3.7.1 On-Chip Silicon Lasers

Since silicon is an indirect-bandgap material, it is not naturally capable of achieving efficient radiative recombination required for producing a high-quality laser. However, silicon can potentially serve as a platform for lasing by incorporating other gain materials [91]. As these emerging technologies continue to mature, and high-quality on-chip lasers compatible with CMOS fabrication techniques come to fruition, these devices will fill a critical gap that currently exists in systems based on silicon photonics.

Recent efforts in epitaxial growth of germanium on silicon have used tensile strain and heavy doping to produce a nearly direct bandgap, and room-temperature optically-pumped lasing has been demonstrated [92]. Lasers based on doped and strained germanium, especially those that are electrically pumped [93], have great potential of rounding out the utility of silicon-based photonics. Other promising techniques have produced electrically-pumped hybrid silicon lasers, where the optical mode is evanescently coupled to III-V compound semiconductors [94]. Furthermore, other research efforts have aimed at producing electrically-pumped rare-earth-ion lasers on silicon [95, 96].

3.7.2 Off-Chip Compound Semiconductor Lasers

More near-term solutions leverage III-V compound semiconductors to produce efficient external off-chip lasers, and couple the produced light to the silicon chip. Quantum dot lasers, based on III-V compound semiconductor quantum dots, are capable of producing many narrow-spectrum peaks over an entire wavelength band that can be used for WDM applications [97]. Coupled with broadband quantum dot semiconductor optical amplifiers, these lasers are capable of producing many wavelength channels, with low relative intensity noise, that may be modulated, transmitted, and received with error-free performance [98]. This approach alleviates packaging complexity of the silicon chip, while simultaneously minimizing the overall cost and power consumption. Another potential approach is to use parametric oscillation in silicon nitride microring resonators to convert an off-chip single-wavelength source to an on-chip multiple-wavelength WDM comb [99].

3.8 Couplers

Couplers are critical silicon photonic building blocks that allow on-chip devices to physically interface with off-chip components. In most silicon photonic applications, the high index contrast between the silicon core and the silicon dioxide cladding results in an extremely small optical mode size for single-mode operation. Although this condition enables very dense integration of silicon photonic devices, it also makes it challenging to efficiently couple between the optical mode of the waveguides (with about 0.3-μm $\times$ 0.3-μm cross-sections) and single-mode fibers (with about 10-μm diameters), producing about a 1,000$\times$ mismatch in the cross-sectional area of the optical mode.

Several key challenges exist in developing adequate coupling techniques, including insertion loss, integration density, bandwidth density, crosstalk, reflectivity, and scalability. There are two promising coupling methods that have been developed for this purpose, using lateral [100–103] and vertical [104–106] techniques.

Another major challenge for each coupling technology, critical for its commercial viability, is its packaging. Each demonstrated coupling method is further accompanied by corresponding optical packaging techniques.

3.8.1 Lateral Couplers

Lateral coupling may prove to be the most feasible solution with 3D integration, since it doesn't depend on the availability of the chip surface. There have been several demonstrations that have leveraged the inverse-taper techniques to expand the optical mode for efficient lateral coupling [100–103, 107, 108]. In some demonstrations, the inverse tapers were also evanescently coupled to polymer waveguides to further expand and guide the optical mode [100, 102], which may not be compatible with CMOS processing.

3.8.1.1 Packaging for Lateral Couplers

One promising packaging demonstration has leveraged CMOS-compatible SiON to create efficient spot-size converting waveguide couplers that were used in a multi-channel coupling to a single-mode fiber array [103]. This experimental demonstration has achieved a 1-dB coupling loss, −35-dB crosstalk with eight-channel coupling. During a four-channel demonstration, 8-Tb/s/mm bandwidth densities were verified with a 20-μm pitch and 160-Gb/s bandwidths for each channel. This work leveraged active alignment to achieve this multi-channel coupling.

Another demonstration featured an inverse-tapered crystalline silicon waveguide fabricated with a CMOS-compatible V-groove structure, enabling passive alignment between the on-chip waveguide and an off-chip single-mode fiber [108]. The V-groove structure, which is fabricated using stable and predictable etching strategies, enables very accurate placement of the fiber, whose optical mode is set to align well with the optical mode of the waveguide. Experimentally, the fabricated structure produced a 7.5-dB coupling loss, spanning over 70 nm of the spectrum, which may eventually be reduced to about 1 dB with improved optical mode matching.

To bridge the packaging gap using lateral couplers, the two aforementioned empirical studies may be combined to produce a CMOS-compatible solution that is both densely-integrated, with adequate scalability, and passively aligned, with suitable mechanical stability.

Analogous to wire bonding of electrical ICs, photonic wire bonding (PWB) was demonstrated to optically interconnect silicon photonic chips [109]. This method utilizes 3D in-situ structuring of negative-tone resist materials with two-photon polymerization, which does not require high-precision alignment of photonic devices, nor is it inherently limited by the spatial density of optical fiber. Simulations for single-mode PWB interconnecting two on-chip inverse-tapered structures yield transmission efficiencies up to 95 %. The pitch between these interconnects can be smaller than 5 μm, corresponding to a density of tens of thousands of these interconnects per square millimeter.

3.8.2 Vertical Couplers

Vertical coupling may prove to be the most efficient for mass production since it allows wafer-scale testing. This means that the silicon photonic devices can be probed to verify and quantify the functionalities before the waver is diced into chips. Efficient vertical couplers have been proposed and demonstrated using various types of grating structures [104–106]. Utilizing a lithographically-defined holographic lens, recent efforts have demonstrated the capability of efficiently coupling light from normal incidence to the chip [106]. These holographic lenses experimentally produced insertion losses of about 1.5 dB between 1530 and 1560 nm, with the best simulated insertion losses of around 0.3 dB.

Compared to lateral couplers, vertical couplers may also have more relaxed alignment tolerances. The coupling tolerances for a 1-dB insertion loss penalty are between $\pm 0.4\,\mu m$ and $\pm 0.7\,\mu m$ for lateral couplers [103], and between $\pm 1\,\mu m$ and $\pm 2\mu m$ for vertical couplers [110].

3.8.2.1 Packaging for Vertical Couplers

Using grating structures for vertical coupling, there have been several successful packaging demonstrations, including some with commercial viability. One demonstration connected a commercial fiber array to a silicon photonic chip with six grating coupler ports [110]. Alignment sensitivity was verified for all the ports, and the position of the chip was fixed to the fiber array using a ultraviolet-curing epoxy.

Another packaging concept, named *g-Pack*, is a more complete packaging solution that features low-frequency electrical connections, wire bonded to a commercial pin-grid array (PGA) carrier, as well as an off-chip fiber array that aligns to the on-chip grating couplers with a standardized geometry [111].

3.9 Links

Leveraging many of the silicon photonic building blocks already developed for on- and off-chip communication, there have been several key demonstrations of transmission links that offer a glimpse of the integration capability available for building larger-scale photonic systems [32, 112, 113]. These demonstrations have combined many key silicon photonic building blocks to form high-performance photonic transmission links carrying out more complex functionality in a cohesive system.

One of the first demonstrations of a fully-integrated photonic transmission link combined a crystalline silicon microring resonator electro-optic modulator coupled to a crystalline silicon waveguide, which then fed into a germanium photodetector [112, 113]. This photonic link had electrical signaling at the input, feeding the modulator, as well as electrical signaling at the output, being fed by the photodetector. The

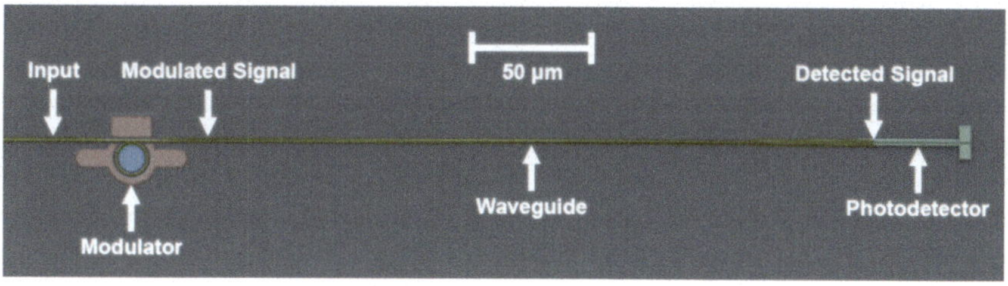

Fig. 3.27 Scanning-Electron-Microscope Image of Silicon Photonic Link [113]: *Top-view* scanning-electron-microscope (SEM) image of the silicon photonic link comprising a crystalline silicon microring resonator electro-optic modulator, crystalline silicon waveguide, and germanium photodetector

modulator was also fed a CW light source generated using an external laser. As shown in Fig. 3.27, these transceivers were integrated together on a single chip, with a high-bandwidth photonic transmission link in between.

As shown in Fig. 3.27, the microring resonator of the modulator had a 12-µm diameter, with a lateral PIN junction across the waveguide [112]. The photodetector was about 40-µm long, and consisted of a waveguide-integrated metal-semiconductor-metal (MSM) device, which enables ultra-low capacitances, as low as 2.4 fF, and fast response times, as low as 8.8 ps [87]. The responsivity of this photodetector was estimated to be as high as 0.9 A/W. The footprints of the modulator and photodetector were about 115 μm^2 and 125 μm^2, respectively, and the silicon photonic waveguide between the two devices is about 250-µm long.

A 3-Gb/s operation of the link was first demonstrated with a total energy consumption of 120 fJ/bit [112]. This link functioned with a 0.5-V voltage swing on the modulator and a 1-V voltage bias on the photodetector, making this link compatible with direct CMOS voltage signaling. In this configuration, open output eye diagrams were observed.

A more recent experimental demonstration using this photonic link has included measuring error-free 3-Gb/s operation [113]. A 2-dB power penalty then was measured for the entire transmission link, including all of its electrical and photonic components.

A further demonstration of silicon photonic systems with multiple materials included low-loss silicon nitride waveguides, polymer-based electro-optic modulators, and polycrystalline germanium photodetectors, also fabricated using CMOS-compatible processes [32].

3.10 Fabrication

The successful emergence of scalable silicon photonic technologies relies on precise fabrication and manufacturability. Fabrication uniformity at the device, chip, wafer, and batch scales determines the performance variation of each type of device. Given the inherent spatial variations that exist in CMOS materials and fabrication processes, the performance impact on silicon photonic devices will be greater than that on electronic devices, especially when using resonant structures [41]. Inherent impairments such as discretization of devices at different locations of the mask reticle, variations in etch depths across a wafer, and batch-to-batch changes in the thickness of the top silicon layer of an SOI wafer, will all have an impact on the uniformity, repeatability, and performance of the silicon photonic devices [41]. For example, the change in height of a standard silicon photonic waveguide forming a microring resonator causes the resonance to be shifted approximately 160 GHz/nm, a significant amount even for fractions of a nanometer. Therefore, some degree of trimming and thermal tuning will be required to make up for the fabrication imperfections, adding cost, complexity, and power dissipation to the systems.

3.11 Integration

All the required functionalities of silicon photonics for photonic interconnection networks—such as waveguides, modulators, switches, photodetectors, lasers and couplers—have already been demonstrated. Importantly, silicon photonics offers compatibility with standard CMOS fabrication processes, enabling dense integration with advanced microelectronics. The capability of silicon photonic devices to be integrated in this highly refined platform, with decades of high-quality development driven by the microprocessor industry, allows the community to think in terms of low-cost mass-volume production.

The exact method of photonic-electronic integration depends on several key factors, including the maximum achievable density and yield of the physical connections, as well as the inherent parasitic capacitance and its resulting power dissipation added with the integration [114]. The reduction of this parasitic capacitance directly improves the receiver sensitivity, improving the scalability, robustness, and energy efficiency of the entire system. Several research groups, including those at Sandia National Laboratories and IBM, have already demonstrated proof-of-concept wire bonding methods to connect electrical chips with advanced CMOS circuits to optical chips with silicon photonic links. Such integration yielded sub-picojoule-per-bit operation of these links, with further power gains being obtained through process refinement and device design.

Research efforts at Oracle and Kotura have leveraged 3D microbumps to bond chips and wafers to each other, offering a method to integrate these inherently heterogeneous systems. Using these microbumps, wafers with high-speed electronics can be integrated with state-of-the-art silicon photonics, with a moderately low capacitance of 20 fF, a pitch of 25 μm, and a yield greater than 99.99 %. This method allows the individual optimization of each unique technology before the bonding.

Other research efforts at MIT are attempting ultralow-capacitance integration of advanced electronics with silicon photonics using the 3D through-oxide via (TOV) method developed by MIT Lincoln Laboratory. The processes for fabricating these devices has been developed and refined to achieve low-risk, 99.999 %-yield, 3-μm-pitch 3D integration with 1-fF capacitance. This directly translates to more-energy-efficient lasers, modulators, and switches, as well as high-sensitivity photodetectors.

Being able to independently fabricate and test the electrical and optical chips before the monolithic integration represents a best-of-both-worlds approach, where only the best-performing devices are merged together. This has the potential to increase the yield and performance of the combined systems. However, these benefits are traded off with cost and complexity, since they require complex physical integration of these inherently heterogeneous systems.

Research efforts at Luxtera, Sandia National Laboratories, MIT, and IBM have leveraged direct front-end integration of silicon photonics with CMOS on the same exact platform. This direct integration of advanced electronics with silicon photonics offers a way to integrate both technologies in a single monolithic silicon system. This method potentially offers record-low capacitances of about 1 fF, high-density integration with a 1-μm pitch, and a near-perfect yield similar to that of transistors in an advanced CMOS integration process. Since this approach simultaneously fabricates both technologies in a single platform, the silicon photonics cannot be optimized separately and brought together with the advanced electronics.

Single-crystalline silicon consists of a perfect lattice of silicon atoms, and generally offers the best set of electrical and optical properties for silicon photonics. However, crystalline silicon can only be grown from another silicon crystal, making it impossible to deposit on non-crystalline substrates, typically limiting the optical devices to a single layer. Polycrystalline silicon is capable of being deposited as an electrically conductive light-guiding material. Polycrystalline silicon does not have a homogeneous crystalline structure, but instead consists of crystalline grains separated by thin disordered grain boundaries. Although polycrystalline silicon may be used for electrically active devices, its propagation losses are intolerable over centimeter-scale distances. Alternatively, silicon nitride is capable of being deposited, with demonstrated low-propagation losses. Although silicon nitride is not useful for making electrically active devices, its relatively high refractive index still allows for high confinement sub-micrometer waveguides. These material systems, which are capable of being deposited together to form three-dimensionally integrated silicon photonic devices, introduce another axis for maximizing the performance of the photonic interconnection networks.

Research efforts at Cornell University and Columbia University have envisioned direct back-end integration of CMOS electronics with these three-dimensionally integrated silicon photonics. This vision leverages silicon photonic materials that are capable of being deposited at the relatively low temperatures compatible with back-end integration processes.

The aforementioned research efforts and demonstrations at leading academic and industry organizations have paved the way for several feasible integration methods to emerge between the silicon photonics and advanced CMOS electronics. These systems will attempt to carefully optimize integration density, yield, parasitic power, and complexity; all of which will also contribute to the overall system feasibility and performance.

References

1. R. A. Soref and B. R. Bennett, "Electrooptical effects in silicon," *Quantum Electronics, IEEE Journal of*, vol. 23, no. 1, pp. 123–129, 1987.
2. S. Manipatruni, Q. Xu, B. Schmidt, J. Shakya, and M. Lipson, "High speed carrier injection 18 Gb/s silicon micro-ring electro-optic modulator," in *Lasers and Electro-Optics Society, 2007. LEOS 2007. The 20th Annual Meeting of the IEEE*, 2007, pp. 537–538.
3. H. L. R. Lira, S. Manipatruni, and M. Lipson, "Broadband hitless silicon electro-optic switch for on-chip optical networks," *Opt. Express*, vol. 17, no. 25, pp. 22 271–22 280, 2009.
4. K. K. Lee, D. R. Lim, L. C. Kimerling, J. Shin, and F. Cerrina, "Fabrication of ultralow-loss Si/SiO2 waveguides by roughness reduction," *Opt. Lett.*, vol. 26, no. 23, pp. 1888–1890, Dec. 2001.
5. F. Y. Gardes, G. T. Reed, A. P. Knights, G. Mashanovich, P. E. Jessop, L. Rowe, S. McFaul, D. Bruce, and N. G. Tarr, "Sub-micron optical waveguides for silicon photonics formed via the local oxidation of silicon (LOCOS)," *Proc. SPIE*, pp. 68 980R–68 980R-4, 2008.
6. P. Dong, W. Qian, S. Liao, H. Liang, C.-C. Kung, N.-N. Feng, R. Shafiiha, J. Fong, D. Feng, A. V. Krishnamoorthy, and M. Asghari, "Low loss shallow-ridge silicon waveguides," *Opt. Express*, vol. 18, no. 14, pp. 14 474–14 479, Jul. 2010.
7. W. Bogaerts and S. Selvaraja, "Compact single-mode silicon hybrid rib/strip waveguide with adiabatic bends," *Photonics Journal, IEEE*, vol. 3, no. 3, pp. 422–432, Jun. 2011.
8. M. Borselli, T. Johnson, and O. Painter, "Beyond the rayleigh scattering limit in high-Q silicon microdisks: theory and experiment," *Opt. Express*, vol. 13, no. 5, pp. 1515–1530, Mar. 2005.
9. G. Li, J. Yao, H. Thacker, A. Mekis, X. Zheng, I. Shubin, Y. Luo, J. hyoung Lee, K. Raj, J. E. Cunningham, and A. V. Krishnamoorthy, "Ultralow-loss, high-density SOI optical waveguide routing for macrochip interconnects," *Opt. Express*, vol. 20, no. 11, pp. 12 035–12 039, May 2012.
10. A. Biberman, M. J. Shaw, E. Timurdogan, J. B. Wright, and M. R. Watts, "Ultralow-loss silicon ring resonators," *Opt. Lett.*, vol. 37, no. 20, pp. 4236–4238, Oct. 2012.
11. A. Biberman, M. Shaw, E. Timurdogan, J. Wright, and M. Watts, "Ultralow-loss silicon ring resonators," in *Group IV Photonics (GFP), 2012 IEEE 9th International Conference on*, Aug. 2012, pp. 39–41.
12. Q. Fang, J. F. Song, S. H. Tao, M. B. Yu, G. Q. Lo, and D. L. Kwong, "Low loss (6.45 dB/cm) sub-micron polycrystalline silicon waveguide integrated with efficient SiON waveguide coupler," *Opt. Express*, vol. 16, no. 9, pp. 6425–6432, 2008.
13. K. Preston, S. Manipatruni, A. Gondarenko, C. B. Poitras, and M. Lipson, "Deposited silicon high-speed integrated electro-optic modulator," *Opt. Express*, vol. 17, no. 7, pp. 5118–5124, 2009.
14. K. Preston, Y. H. D. Lee, M. Zhang, and M. Lipson, "Waveguide-integrated telecom-wavelength photodiode in deposited silicon," *Opt. Lett.*, vol. 36, no. 1, pp. 52–54, Jan. 2011.
15. J. F. Bauters, M. J. R. Heck, D. D. John, J. S. Barton, C. M. Bruinink, A. Leinse, R. G. Heideman, D. J. Blumenthal, and J. E. Bowers, "Planar waveguides with less than 0.1 dB/m propagation loss fabricated with wafer bonding," *Opt. Express*, vol. 19, no. 24, pp. 24 090–24 101, Nov. 2011.
16. M. J. Shaw, J. Guo, G. A. Vawter, S. Habermehl, and C. T. Sullivan, "Fabrication techniques for low-loss silicon nitride waveguides," in *Micromachining Technology for Micro-Optics and Nano-Optics III*, E. G. Johnson, G. P. Nordin, and T. J. Suleski, Eds., vol. 5720, 2005, pp. 109–118.
17. A. Gondarenko, J. S. Levy, and M. Lipson, "High confinement micron-scale silicon nitride high Q ring resonator," *Opt. Express*, vol. 17, no. 14, pp. 11 366–11 370, 2009.
18. S. Zhu, G. Q. Lo, and D. L. Kwong, "Low-loss amorphous silicon wire waveguide for integrated photonics: effect of fabrication process and the thermal stability," *Opt. Express*, vol. 18, no. 24, pp. 25 283–25 291, Nov. 2010.
19. J. Michel, J. Liu, and L. C. Kimerling, "High-performance Ge-on-Si photodetectors," *Nature Photonics*, vol. 4, pp. 527–534, Aug. 2010.

20. L. Colace, G. Altieri, and G. Assanto, "Waveguide photodetectors for the near-infrared in polycrystalline germanium on silicon," *Photonics Technology Letters, IEEE,* vol. 18, no. 9, pp. 1094–1096, May 2006.
21. H. Lee, T. Chen, J. Li, O. Painter, and K. J. Vahala, "Ultra-low-loss optical delay line on a silicon chip," *Nature Communications,* vol. 3, pp. 1–7, May 2012.
22. H. Lee, T. Chen, J. Li, K. Y. Yang, S. Jeon, O. Painter, and K. J. Vahala, "Chemically etched ultrahigh-Q wedge-resonator on a silicon chip," *Nature Photonics,* vol. 6, pp. 369–373, May 2012.
23. R. Adar, M. Serbin, and V. Mizrahi, "Less than 1 dB per meter propagation loss of silica waveguides measured using a ring resonator," *Lightwave Technology, Journal of,* vol. 12, no. 8, pp. 1369–1372, Aug. 1994.
24. A. Biberman, K. Preston, G. Hendry, N. Sherwood-Droz, J. Chan, J. S. Levy, M. Lipson, and K. Bergman, "Photonic network-on-chip architectures using multi-layer deposited silicon materials for high-performance chip multiprocessors," *J. Emerg. Technol. Comput. Syst.,* vol. 7, Jun. 2011.
25. M. Lipson, "Guiding, modulating, and emitting light on silicon-challenges and opportunities," *Lightwave Technology, Journal of,* vol. 23, no. 12, pp. 4222–4238, Dec. 2005.
26. B. Lee, X. Chen, A. Biberman, X. Liu, I.-W. Hsieh, C.-Y. Chou, J. Dadap, F. Xia, W. Green, L. Sekaric, Y. Vlasov, R. Osgood, and K. Bergman, "Ultrahigh-bandwidth silicon photonic nanowire waveguides for on-chip networks," *Photonics Technology Letters,* IEEE, vol. 20, no. 6, pp. 398–400, Mar. 2008.
27. C. Manolatou, S. G. Johnson, S. Fan, P. R. Villeneuve, H. A. Haus, and J. D. Joannopoulos, "High-density integrated optics," *J. Lightwave Technol.,* vol. 17, no. 9, p. 1682, Sep. 1999.
28. W. Bogaerts, P. Dumon, D. V. Thourhout, and R. Baets, "Low-loss, low-cross-talk crossings for silicon-on-insulator nanophotonic waveguides," *Opt. Lett.,* vol. 32, no. 19, pp. 2801–2803, 2007.
29. F. Xu and A. W. Poon, "Silicon cross-connect filters using microring resonator coupled multimode-interference-based waveguide crossings," *Opt. Express,* vol. 16, no. 12, pp. 8649–8657, 2008.
30. M. Popovic, E. Ippen, and F. Kartner, "Low-loss bloch waves in open structures and highly compact, efficient Si waveguide-crossing arrays," in *Lasers and Electro-Optics Society, 2007. LEOS 2007. The 20th Annual Meeting of the IEEE,* Oct. 2007, pp. 56–57.
31. J. Chan, G. Hendry, A. Biberman, and K. Bergman, "Architectural exploration of chip-scale photonic interconnection network designs using physical-layer analysis," *Lightwave Technology, Journal of,* vol. 28, no. 9, pp. 1305–1315, May 2010.
32. I. Young, E. Mohammed, J. Liao, A. Kern, S. Palermo, B. Block, M. Reshotko, and P. Chang, "Optical I/O technology for tera-scale computing," *Solid-State Circuits, IEEE Journal of,* vol. 45, no. 1, pp. 235–248, Jan. 2010.
33. N. Ophir, A. Biberman, J. S. Levy, K. Padmaraju, K. J. Luke, M. Lipson, and K. Bergman, "Demonstration of 1.28-Tb/s transmission in next-generation nanowires for photonic networks-on-chip," in *IEEE Photonics Society, 2010 23rd Annual Meeting of the,* Nov. 2010, pp. 560–561.
34. K. Ikeda, R. E. Saperstein, N. Alic, and Y. Fainman, "Thermal and Kerr nonlinear properties of plasma-deposited silicon nitride/silicon dioxide waveguides," *Opt. Express,* vol. 16, no. 17, pp. 12 987–12 994, Aug. 2008.
35. A. Biberman and K. Bergman, "Optical interconnection networks for high-performance computing systems," *Reports on Progress in Physics,* vol. 75, no. 4, 2012.
36. A. Krishnamoorthy, R. Ho, X. Zheng, H. Schwetman, J. Lexau, P. Koka, G. Li, I. Shubin, and J. Cunningham, "Computer systems based on silicon photonic interconnects," *Proceedings of the IEEE,* vol. 97, no. 7, pp. 1337–1361, Jul. 2009.
37. C. Batten, A. Joshi, J. Orcutt, A. Khilo, B. Moss, C. Holzwarth, M. Popovic, H. Li, H. Smith, J. Hoyt, F. Kartner, R. Ram, V. Stojanovic, and K. Asanovic, "Building many-core processor-to-DRAM networks with monolithic CMOS silicon photonics," *Micro, IEEE,* vol. 29, no. 4, pp. 8–21, Jul.-Aug. 2009.

38. R. Beausoleil, J. Ahn, N. Binkert, A. Davis, D. Fattal, M. Fiorentino, N. P. Jouppi, M. McLaren, C. M. Santori, R. S. Schreiber, S. M. Spillane, D. Vantrease, and Q. Xu, "A nanophotonic interconnect for high-performance many-core computation," in *Integrated Photonics and Nanophotonics Research and Applications*, 2008, p. ITuD2.
39. A. Shacham, K. Bergman, and L. P. Carloni, "On the design of a photonic network-on-chip," in *NOCS '07: Proceedings of the First International Symposium on Networks-on-Chip,* 2007, pp. 53–64.
40. N. Kirman, M. Kirman, R. K. Dokania, J. F. Martinez, A. B. Apsel, M. A. Watkins, and D. H. Albonesi, "Leveraging optical technology in future bus-based chip multiprocessors," in *MICRO 39: Proceedings of the 39th Annual IEEE/ACM International Symposium on Microarchitecture,* 2006, pp. 492–503.
41. R. G. Beausoleil, "Large-scale integrated photonics for high-performance interconnects," *J. Emerg. Technol. Comput. Syst.,* vol. 7, no. 2, pp. 6:1–6:54, Jul. 2011.
42. Q. Xu, D. Fattal, and R. G. Beausoleil, "Silicon microring resonators with 1.5-μm radius," *Opt. Express,* vol. 16, no. 6, pp. 4309–4315, 2008.
43. S. Manipatruni, K. Preston, L. Chen, M. Lipson, "Ultra-low voltage, ultra-small mode volume silicon microring modulator," in *Opt. Express,* vol. 18, 2010, pp. 18 235–18 242.
44. W. A. Zortman, M. R. Watts, D. C. Trotter, R. W. Young, and A. L. Lentine, "Low-power high-speed silicon microdisk modulators," in *Conference on Lasers and Electro-Optics,* 2010, p. CThJ4.
45. Q. Xu, B. Schmidt, S. Pradhan, and M. Lipson, "Micrometre-scale silicon electro-optic modulator," *Nature,* vol. 435, pp. 325–327, 05 2005.
46. X. Xiao, H. Xu, X. Li, Y. Hu, K. Xiong, Z. Li, T. Chu, Y. Yu, and J. Yu, "25 Gbit/s silicon microring modulator based on misalignment-tolerant interleaved PN junctions," *Opt. Express,* vol. 20, no. 3, pp. 2507–2515, Jan. 2012.
47. J. C. Rosenberg, W. M. J. Green, S. Assefa, D. M. Gill, T. Barwicz, M. Yang, S. M. Shank, and Y. A. Vlasov, "A 25 Gbps silicon microring modulator based on an interleaved junction," *Opt. Express,* vol. 20, no. 24, pp. 26 411–26 423, Nov. 2012.
48. G. Li, X. Zheng, J. Yao, H. Thacker, I. Shubin, Y. Luo, K. Raj, J. E. Cunningham, and A. V. Krishnamoorthy, "25 Gb/s 1 V-driving CMOS ring modulator with integrated thermal tuning," *Opt. Express,* vol. 19, no. 21, pp. 20 435–20 443, Oct. 2011.
49. G. Li, X. Zheng, H. Thacker, J. Yao, Y. Luo, I. Shubin, K. Raj, J. Cunningham, and A. Krishnamoorthy, "40 Gb/s thermally tunable CMOS ring modulator," in *Group IV Photonics (GFP), 2012 IEEE 9th International Conference on,* Aug. 2012, pp. 1–3.
50. Y. Hu, X. Xiao, H. Xu, X. Li, K. Xiong, Z. Li, T. Chu, Y. Yu, and J. Yu, "High-speed silicon modulator based on cascaded microring resonators," *Opt. Express,* vol. 20, no. 14, pp. 15 079–15 085, Jul. 2012.
51. X. Xiao, X. Li, H. Xu, Y. Hu, K. Xiong, Z. Li, T. Chu, J. Yu, and Y. Yu, "44-Gb/s silicon microring modulators based on zigzag PN junctions," *Photonics Technology Letters, IEEE,* vol. 24, no. 19, pp. 1712–1714, Oct. 2012.
52. M. R. Watts, W. A. Zortman, D. C. Trotter, R. W. Young, and A. L. Lentine, "Vertical junction silicon microdisk modulators and switches," *Opt. Express,* vol. 19, no. 22, pp. 21 989–22 003, Oct. 2011.
53. A. Biberman, E. Timurdogan, W. A. Zortman, D. C. Trotter, and M. R. Watts, "Adiabatic microring modulators," *Opt. Express,* vol. 20, no. 28, pp. 29 223–29 236, Dec. 2012.
54. E. Timurdogan, M. Moresco, A. Biberman, J. Sun, W. Zortman, D. Trotter, and M. Watts, "Adiabatic resonant microring (ARM) modulator," in *Optical Interconnects Conference, 2012 IEEE,* May 2012, pp. 48–49.
55. M. R. Watts, "Adiabatic microring resonators," *Opt. Lett.,* vol. 35, no. 19, pp. 3231–3233, Oct. 2010.
56. M. Watts, W. Zortman, D. Trotter, G. Nielson, D. Luck, and R. Young, "Adiabatic resonant microrings (ARMs) with directly integrated thermal microphotonics," in *Lasers and Electro-Optics, 2009 and 2009 Conference on Quantum electronics and Laser Science Conference. CLEO/QELS 2009. Conference on,* Jun. 2009, pp. 1–2.

57. A. Biberman, S. Manipatruni, N. Ophir, L. Chen, M. Lipson, and K. Bergman, "First demonstration of long-haul transmission using silicon microring modulators," *Opt. Express,* vol. 18, no. 15, pp. 15 544–15 552, Jul. 2010.
58. A. Biberman, N. Ophir, K. Bergman, S. Manipatruni, L. Chen, and M. Lipson, "First experimental bit-error-rate validation of 12.5-Gb/s silicon modulator enabling photonic networks-on-chip," in *Optical Fiber Communication Conference,* 2010, p. OMI1.
59. A. Biberman, N. Ophir, K. Bergman, S. Manipatruni, L. Chen, and M. Lipson, "First demonstration of 80-km long-haul transmission of 12.5-Gb/s data using silicon microring resonator electro-optic modulator," in *Optical Fiber Communication Conference,* 2010, p. JWA28.
60. S. Manipatruni, L. Chen, and M. Lipson, "50 Gbit/s wavelength division multiplexing using silicon microring modulators," in *Group IV Photonics, 2009. GFP '09. 6th IEEE International Conference on,* 2009, pp. 244–246.
61. Q. Xu, B. Schmidt, J. Shakya, and M. Lipson, "Cascaded silicon micro-ring modulators for WDM optical interconnection," *Opt. Express,* vol. 14, no. 20, pp. 9431–9435, Oct. 2006.
62. B. Lee, B. Small, Q. Xu, M. Lipson, and K. Bergman, "Characterization of a 4 × 4 Gb/s parallel electronic bus to WDM optical link silicon photonic translator," *Photonics Technology Letters, IEEE,* vol. 19, no. 7, pp. 456–458, Apr. 2007.
63. K. Padmaraju, N. Ophir, A. Biberman, L. Chen, E. Swan, J. Chan, M. Lipson, and K. Bergman, "Intermodulation crosstalk from silicon microring modulators in wavelength-parallel photonic networks-on-chip," in *IEEE Photonics Society, 2010 23rd Annual Meeting of the,* Nov. 2010, pp. 562–563.
64. A. Yariv, "Universal relations for coupling of optical power between microresonators and dielectric waveguides," *Electronics Letters,* vol. 36, no. 4, pp. 321–322, Feb. 2000.
65. J. T. Robinson, K. Preston, O. Painter, and M. Lipson, "First-principle derivation of gain in high-index-contrast waveguides," *Opt. Express,* vol. 16, no. 21, pp. 16 659–16 669, Oct. 2008.
66. M. R. Watts, D. C. Trotter, and R. W. Young, "Maximally confined high-speed second-order silicon microdisk switches," in *National Fiber Optic Engineers Conference,* 2008, p. PDP14.
67. A. Biberman, H. L. Lira, K. Padmaraju, N. Ophir, M. Lipson, and K. Bergman, "Broadband CMOS-compatible silicon photonic electro-optic switch for photonic networks-on-chip," in *Conference on Lasers and Electro-Optics,* 2010, p. CPDA11.
68. Y. Vlasov, W. M. J. Green, and F. Xia, "High-throughput silicon nanophotonic wavelength-insensitive switch for on-chip optical networks," *Nature Photonics,* vol. 2, pp. 242–246, Apr. 2008.
69. A. Biberman, H. Lira, K. Padmaraju, N. Ophir, J. Chan, M. Lipson, and K. Bergman, "Broadband silicon photonic electrooptic switch for photonic interconnection networks," *Photonics Technology Letters, IEEE,* vol. 23, no. 8, pp. 504–506, Apr. 2011.
70. B. Lee, A. Biberman, P. Dong, M. Lipson, and K. Bergman, "All-optical comb switch for multiwavelength message routing in silicon photonic networks," *Photonics Technology Letters, IEEE,* vol. 20, no. 10, pp. 767–769, May 2008.
71. B. Lee, A. Biberman, N. Sherwood-Droz, C. Poitras, M. Lipson, and K. Bergman, "High-speed 2 × 2 switch for multi-wavelength message routing in on-chip silicon photonic networks," in *Optical Communication, 2008. ECOC 2008. 34th European Conference on,* Sep. 2008, pp. 1–2.
72. A. Biberman, B. G. Lee, N. Sherwood-Droz, M. Lipson, and K. Bergman, "Broadband operation of nanophotonic router for silicon photonic networks-on-chip," *Photonics Technology Letters, IEEE,* vol. 22, no. 12, pp. 926–928, Jun. 2010.
73. A. Biberman, P. Dong, B. Lee, J. Foster, M. Lipson, and K. Bergman, "Silicon microring resonator-based broadband comb switch for wavelength-parallel message routing," in *Lasers and Electro-Optics Society, 2007. LEOS 2007. The 20th Annual Meeting of the IEEE,* Oct. 2007, pp. 474–475.
74. A. Biberman, B. G. Lee, K. Bergman, P. Dong, and M. Lipson, "Demonstration of all-optical multi-wavelength message routing for silicon photonic networks," in *Optical Fiber Communication Conference and Exposition and The National Fiber Optic Engineers Conference,* 2008, p. OTuF6.

75. P. Dong, S. F. Preble, and M. Lipson, "All-optical compact silicon comb switch," *Opt. Express,* vol. 15, no. 15, pp. 9600–9605, 2007.
76. A. Biberman, B. Lee, P. Dong, M. Lipson, and K. Bergman, "250 Gb/s multi-wavelength operation of microring resonator-based broadband comb switch for silicon photonic networks-on-chip," in *Optical Communication, 2008. ECOC 2008. 34th European Conference on,* Sep. 2008, pp. 1–2.
77. B. G. Lee, A. Biberman, N. Sherwood-Droz, C. B. Poitras, M. Lipson, and K. Bergman, "High-speed 2 × 2 switch for multiwavelength silicon-photonic networks-on-chip," *J. Lightwave Technol.,* vol. 27, no. 14, pp. 2900–2907, Jul. 2009.
78. H. Wang, M. Petracca, A. Biberman, B. G. Lee, L. P. Carloni, and K. Bergman, "Nanophotonic optical interconnection network architecture for on-chip and off-chip communications," in *Optical Fiber Communication Conference and Exposition and The National Fiber Optic Engineers Conference,* 2008, p. JThA92.
79. A. Shacham, B. Lee, A. Biberman, K. Bergman, and L. Carloni, "Photonic NoC for DMA communications in chip multiprocessors," in *High-Performance Interconnects, 2007. HOTI 2007. 15th Annual IEEE Symposium on,* Aug. 2007, pp. 29–38.
80. N. Sherwood-Droz, H. Wang, L. Chen, B. G. Lee, A. Biberman, K. Bergman, and M. Lipson, "Optical 4 × 4 hitless silicon router for optical networks-on-chip (NoC)," *Opt. Express,* vol. 16, no. 20, pp. 15 915–15 922, Sep. 2008.
81. B. G. Lee, A. Biberman, K. Bergman, N. Sherwood-Droz, and M. Lipson, "Multi-wavelength message routing in a non-blocking four-port bidirectional switch fabric for silicon photonic networks-on-chip," in *Optical Fiber Communication Conference,* 2009, p. OMJ4.
82. A. Biberman, N. Sherwood-Droz, B. Lee, M. Lipson, and K. Bergman, "Thermally active 4 × 4 non-blocking switch for networks-on-chip," in *IEEE Lasers and Electro-Optics Society, 2008. LEOS 2008. 21st Annual Meeting of the,* Nov. 2008, pp. 370–371.
83. Q. Xu, S. Manipatruni, B. Schmidt, J. Shakya, and M. Lipson, "12.5 Gbit/s carrier-injection-based silicon micro-ring silicon modulators," *Opt. Express,* vol. 15, no. 2, pp. 430–436, Jan. 2007.
84. J. Chan, A. Biberman, B. G. Lee, and K. Bergman, "Insertion loss analysis in a photonic interconnection network for on-chip and off-chip communications," in *IEEE Lasers and Electro-Optics Society (LEOS),* Nov. 2008.
85. L. Vivien, J. Osmond, J.-M. Fédéli, D. Marris-Morini, P. Crozat, J.-F. Damlencourt, E. Cassan, Y. Lecunff, and S. Laval, "42 GHz P.I.N germanium photodetector integrated in a silicon-on-insulator waveguide," *Opt. Express,* vol. 17, no. 8, pp. 6252–6257, 2009.
86. S. Assefa, F. Xia, S. W. Bedell, Y. Zhang, T. Topuria, P. M. Rice, and Y. A. Vlasov, "CMOS-integrated high-speed MSM germanium waveguide photodetector," *Opt. Express,* vol. 18, no. 5, pp. 4986–4999, 2010.
87. L. Chen and M. Lipson, "Ultra-low capacitance and high speed germanium photodetectors on silicon," *Opt. Express,* vol. 17, no. 10, pp. 7901–7906, 2009.
88. T. Yin, R. Cohen, M. M. Morse, G. Sarid, Y. Chetrit, D. Rubin, and M. J. Paniccia, "31 GHz Ge n-i-p waveguide photodetectors on silicon-on-insulator substrate," *Opt. Express,* vol. 15, no. 21, pp. 13 965–13 971, 2007.
89. J. D. B. Bradley, P. E. Jessop, and A. P. Knights, "Silicon waveguide-integrated optical power monitor with enhanced sensitivity at 1550 nm," *Applied Physics Letters,* vol. 86, no. 24, pp. 241 103–241 103–3, Jun. 2005.
90. M. W. Geis, S. J. Spector, M. E. Grein, J. U. Yoon, D. M. Lennon, and T. M. Lyszczarz, "Silicon waveguide infrared photodiodes with >35 GHz bandwidth and phototransistors with 50 AW^{-1} response," *Opt. Express,* vol. 17, no. 7, pp. 5193–5204, 2009.
91. D. Liang and J. E. Bowers, "Recent progress in lasers on silicon," *Nature Photonics,* vol. 4, pp. 511–517, Aug. 2010.
92. J. Liu, X. Sun, R. Camacho-Aguilera, L. C. Kimerling, and J. Michel, "Ge-on-Si laser operating at room temperature," *Opt. Lett.,* vol. 35, no. 5, pp. 679–681, Mar. 2010.
93. R. E. Camacho-Aguilera, Y. Cai, N. Patel, J. T. Bessette, M. Romagnoli, L. C. Kimerling, and J. Michel, "An electrically pumped germanium laser," *Opt. Express,* vol. 20, no. 10, pp. 11 316–11 320, May 2012.

94. J. Van Campenhout, L. Liu, P. Romeo, D. Van Thourhout, C. Seassal, P. Regreny, L. Di Cioccio, J.-M. Fedeli, and R. Baets, "A compact SOI-integrated multiwavelength laser source based on cascaded InP microdisks," *Photonics Technology Letters, IEEE,* vol. 20, no. 16, pp. 1345–1347, Aug. 2008.
95. O. Jambois, F. Gourbilleau, A. J. Kenyon, J. Montserrat, R. Rizk, and B. Garrido, "Towards population inversion of electrically pumped Er ions sensitized by Si nanoclusters," *Opt. Express,* vol. 18, no. 3, pp. 2230–2235, Feb. 2010.
96. S. Yerci, R. Li, and L. D. Negro, "Electroluminescence from Er-doped Si-rich silicon nitride light emitting diodes," *Applied Physics Letters,* vol. 97, no. 8, p. 081109, 2010.
97. D. Livshits, D. Yin, A. Gubenko, I. Krestnikov, S. Mikhrin, A. Kovsh, and G. Wojcik, "Cost-effective WDM optical interconnects enabled by quantum dot comb lasers," in *Optoelectronic Interconnects and Component Integration IX,* A. L. Glebov and R. T. Chen, Eds., vol. 7607, 2010, p. 76070W.
98. A. Gubenko, I. Krestnikov, D. Livshtis, S. Mikhrin, A. Kovsh, L. West, C. Bornholdt, N. Grote, and A. Zhukov, "Error-free 10 Gbit/s transmission using individual Fabry-Perot modes of low-noise quantum-dot laser," *Electronics Letters,* vol. 43, no. 25, pp. 1430–1431, 6 2007.
99. J. S. Levy, A. Gondarenko, M. A. Foster, A. C. Turner-Foster, A. L. Gaeta, and M. Lipson, "CMOS-compatible multiple-wavelength oscillator for on-chip optical interconnects," *Nature Photonics,* vol. 4, pp. 37–40, Jan. 2010.
100. T. Shoji, T. Tsuchizawa, T. Watanabe, K. Yamada, and H. Morita, "Low loss mode size converter from 0.3 μm square Si wire waveguides to singlemode fibres," *Electronics Letters,* vol. 38, no. 25, pp. 1669–1670, Dec. 2002.
101. V. R. Almeida, R. R. Panepucci, and M. Lipson, "Nanotaper for compact mode conversion," *Opt. Lett.,* vol. 28, no. 15, pp. 1302–1304, Aug. 2003.
102. S. McNab, N. Moll, and Y. Vlasov, "Ultra-low loss photonic integrated circuit with membrane-type photonic crystal waveguides," *Opt. Express,* vol. 11, no. 22, pp. 2927–2939, Nov. 2003.
103. F. E. Doany, B. G. Lee, S. Assefa, W. M. J. Green, M. Yang, C. L. Schow, C. V. Jahnes, S. Zhang, J. Singer, V. I. Kopp, J. A. Kash, and Y. A. Vlasov, "Multichannel high-bandwidth coupling of ultradense silicon photonic waveguide array to standard-pitch fiber array," *J. Lightwave Technol.,* vol. 29, no. 4, pp. 475–482, Feb. 2011.
104. R. Emmons and D. Hall, "Buried-oxide silicon-on-insulator structures. II. Waveguide grating couplers," *Quantum Electronics, IEEE Journal of,* vol. 28, no. 1, pp. 164–175, Jan. 1992.
105. D. Taillaert, H. Chong, P. Borel, L. Frandsen, R. D. L. Rue, and R. Baets, "A compact two-dimensional grating coupler used as a polarization splitter," *Photonics Technology Letters, IEEE,* vol. 15, no. 9, pp. 1249–1251, Sep. 2003.
106. C. Gunn, "CMOS photonics for high-speed interconnects," *Micro, IEEE,* vol. 26, no. 2, pp. 58–66, Mar.-Apr. 2006.
107. L. Chen, C. Doerr, Y.-K. Chen, and T.-Y. Liow, "Low-loss and broadband cantilever couplers between standard cleaved fibers and high-index-contrast Si_3N_4 or Si waveguides," *Photonics Technology Letters, IEEE,* vol. 22, pp. 1744–1746, Dec. 2010.
108. J. Galan, P. Sanchis, J. Marti, S. Marx, H. Schroder, B. Mukhopadhyay, T. Tekin, S. Selvaraja, W. Bogaerts, P. Dumon, and L. Zimmermann, "CMOS compatible silicon etched V-grooves integrated with a SOI fiber coupling technique for enhancing fiber-to-chip alignment," in *Group IV Photonics, 2009. GFP '09. 6th IEEE International Conference on,* Sep. 2009, pp. 148–150.
109. N. Lindenmann, I. Kaiser, G. Balthasar, R. Bonk, D. Hillerkuss, W. Freude, J. Leuthold, and C. Koos, "Photonic waveguide bonds - a novel concept for chip-to-chip interconnects," in *Optical Fiber Communication Conference,* 2011, p. PDPC1.
110. T. Tekin, H. Schroder, L. Zimmermann, P. Dumon, and W. Bogaerts, "Fibre-array optical interconnection for silicon photonics," in *Optical Communication, 2008. ECOC 2008. 34th European Conference on,* Sep. 2008, pp. 1–2.
111. L. Zimmermann, G. Preve, T. Tekin, T. Rosin, and K. Landles, "Packaging and assembly for integrated photonics - a review of the ePIXpack photonics packaging platform," Selected Topics in *Quantum Electronics, IEEE Journal of,* vol. 17, no. 3, pp. 645–651, May-Jun. 2011.

112. L. Chen, K. Preston, S. Manipatruni, and M. Lipson, "Integrated ghz silicon photonic interconnect with micrometer-scale modulators and detectors," *Opt. Express,* vol. 17, no. 17, pp. 15 248–15 256, Aug. 2009.
113. N. Ophir, K. Padmaraju, A. Biberman, L. Chen, K. Preston, M. Lipson, and K. Bergman, "First demonstration of error-free operation of a full silicon on-chip photonic link," in *Optical Fiber Communication Conference,* 2011, p. OWZ3.
114. A. Biberman, "Silicon photonic revolution through advanced integration," *Future Fab International,* pp. 25–28, Jul. 2012.

Chapter 4
Photonic Simulation and Design Space

As discussed in the previous chapter, the progress in silicon photonics research has enabled the physical demonstration of all the devices that are necessary to build extremely high-bandwidth density and energy-efficient links for on-chip and off-chip communications. Photonic network design, however, requires a major paradigm shift from traditional network design due to the fundamental differences in how electronics and photonics operate. Consequently, new modeling and analysis methods must be employed to realize a chip-scale photonic interconnection network. This chapter describes a methodology and a supporting computer-aided design (CAD) environment to model the basic photonic devices, to combine them to realize photonic network architectures, and to analyze the physical-layer and system-level performance properties of these networks.

4.1 Performance Simulation

A methodology for designing, modeling, and analyzing the performance of photonic interconnection networks is described in this section. The methodology encompasses several techniques to synergistically study a photonic architecture's system-level properties through physical-layer analysis. The PhoenixSim design environment was designed to implement the described modeling and analysis aspects of the methodology [1]. PhoenixSim is implemented using OMNeT++, an open-source C++-based event-driven simulation environment [2]. The described methodology and PhoenixSim represent a set of tools which system architects can use to see how integrated photonics can potentially impact the performance of a particular computing system. While PhoenixSim was initially planned and developed targeting chip-scale silicon-photonic architectures, both the underlying methodology and the design environment can be generalized for photonic components based on any material system (e.g. III–V materials) and having any scale (e.g. wide-area networks, telecom).

K. Bergman et al., *Photonic Network-on-Chip Design*,
Integrated Circuits and Systems, DOI: 10.1007/978-1-4419-9335-9_4,

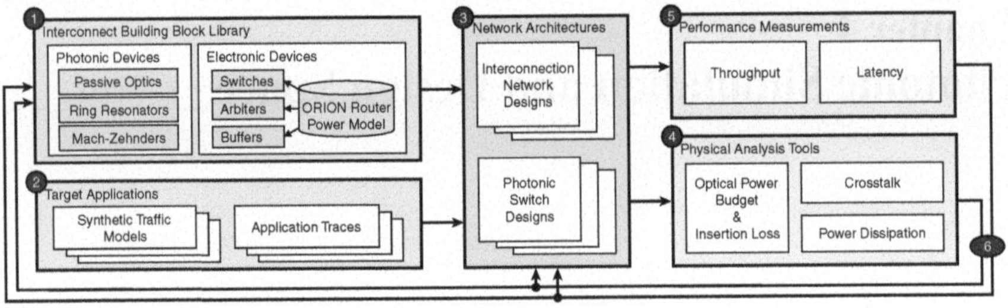

Fig. 4.1 The design flow of modeling a network in the PhoenixSim environment [1]

4.1.1 Motivation for Photonic Simulation

While there are currently a large number of high-quality simulation environments available for studying networks, none are capable of handling the unique network architectures that are possible when considering chip-scale silicon photonics. Notable simulators of traditional systems include ns-2, NetSim, OPNET, and GloMoSim. These simulators typically support most standardized communication protocols (e.g. TCP) and are therefore well suited for traditional large scale networks. While these network simulators may support optical components, the library of elements that they provide are mostly limited to commercially available devices.

An additional simulator characteristic that is needed is the ability to model the physical level of the optical components. Current fabrication technology is limited to simple device-level demonstrations (at most) for silicon photonics. A full scale network exceeds what is reasonably possible in today's fabrication foundries. PhoenixSim is primarily categorized as a system-level simulation environment that includes some aspects of link-level simulation.

4.1.2 Methodology and Design Flow Overview

An overview of the design methodology is illustrated in Fig. 4.1. Modeling of photonic interconnection networks primarily consists of six design steps: (1) specification of the network building blocks, (2) specification of the target application, (3) modeling of the network architecture, (4) system-level performance analysis, (5) physical-layer characterization, and (6) iterative refinement of parameters and design.

Step 1 (as labeled in Fig. 4.1) involves the specification of the fundamental network building blocks that will be used for creating the interconnection network. The collection of network building blocks is named the *Interconnect Building Block Library*. Within this library is a set of photonic devices that are characterized using the Basic Element Device Model (Fig. 4.2), described in further detail in Sect. 4.1.3. Users of this design methodology can choose to design a network based on the

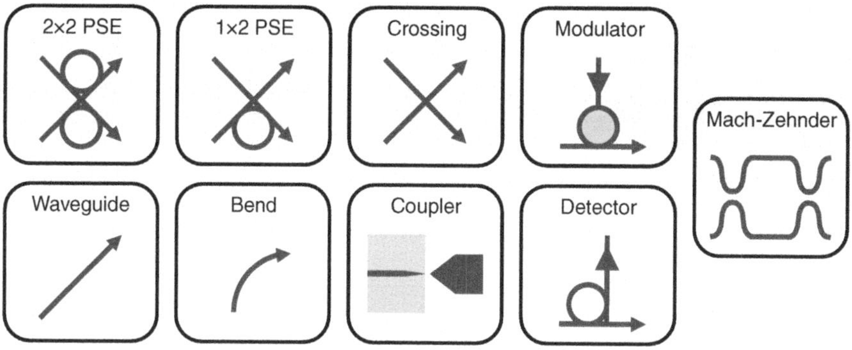

Fig. 4.2 A subset of the photonic devices in the *Interconnect Building Block Library* [1]

included library of devices, or extend the library themselves with other novel photonic building blocks.

The library for electronic building blocks consists of switch, arbitrator, and buffer blocks for creating standard pipelined routers. PhoenixSim leverages the ORION simulator [3] for deriving detailed values for electronic delay and energy dissipation. The electronic router model is highly configurable and includes parameters for clock rate, buffer size, channel width, and number of virtual channels. In addition to the standard router design, the electronic router model also includes additional methods for interfacing with photonic devices. Electro-optic photonic devices can take an electronic input to influence its optical behavior and are essential components for enabling the active types of switching used in some proposed networks [4, 5].

Next, Step 2 consists of specifying the target application. PhoenixSim supports the use of both synthetically generated traffic patterns and communication traces. A variety of synthetic patterns have already been created within the environment (e.g. random, hotspot, nearest neighbor, and tornado) and is extensible to others. Communication traces can be generated by monitoring the network traffic during the execution of a real application and used as an input into PhoenixSim. Performance results gained by using communication traces are useful in assessing the application-specific performance gains of photonic networks [6].

The design and modeling of the network occurs in Step 3 of the design flow. The devices from the *Interconnect Building Block Library* can be combined to create higher-order networking components and entire interconnection network topologies. By accounting for the target applications, a network architect can optimize the topology design to target specific requirements such as message size, latency, and/or throughput. For instance, Fig. 4.3 illustrates how a 4×4 non-blocking switch can be derived within PhoenixSim by connecting various devices from the *Building Block Library*. Figure 4.3a illustrates the schematic representation of the 4×4 non-blocking switch, while Fig. 4.3b depicts the PhoenixSim representation as composed within the environment. In Fig. 4.3c, an image of an actual 4×4 non-blocking switch that was fabricated at the Cornell Nanofabrication Facility is shown [7].

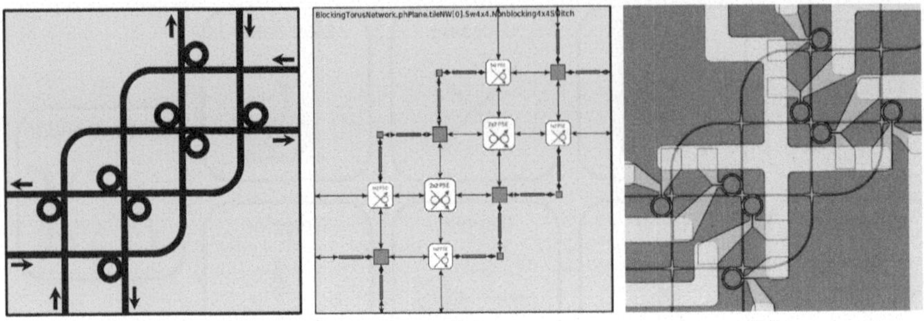

(a) (b) (c)

Fig. 4.3 **a** Schematic of a design for a 4 × 4 non-blocking photonic switch. **b** A screenshot of how PhoenixSim composes the switch by instancing basic photonic devices. **c** Microscope image of a 4 × 4 non-blocking switch fabricated at the Cornell Nanofabrication Facility [1]

Step 4 involves the characterization of the network architecture at the physical-layer, which involves metrics such as the optical power budget, crosstalk, and power dissipation. The overall physical-layer performance of a derived photonic component or topology can be determined from the aggregate performance of the individual photonic devices. Although this is not as rigorous as a true link-level simulator, this hierarchical building process enables an accurate first-order physical characterization of an entire network through the characterization of a small number of foundational components.

Step 5 measures the system-level performance characteristics of the network architecture in terms of data throughput and latency. Many of the physical properties that are identified in Step 4 have an impact on network functionality and scalability and play a crucial role in determining overall system performance.

Finally, Step 6 forms the basis for an iterative process, where the performance results and analysis of the modeled network can be used to refine the topology design and device parameters to further optimize the overall performance. Previous work has demonstrated the effectiveness of this iterative step. The initial physical-layer characterizations showed the dramatic impact that waveguide crossing loss had on performance and a subsequent analysis of a system with improved crossings resulted in a dramatic improvement in overall performance [8].

4.1.3 Photonic Device Library

The PhoenixSim method for modeling photonic devices is designed to enable the assessment of the physical-layer performance at a first-order approximation while concurrently allowing for system-level analysis with a reasonable computational requirement. Many simulation packages use techniques such as finite-difference time-domain (FDTD) to accurately model an electromagnetic field according to Maxwell's equations. FDTD analysis, however, is usually limited to a single or small set of devices since it is computationally intensive and can have a large memory requirement. PhoenixSim uses a more efficient level of abstraction by establishing a set of characteristic device parameters that are key to measuring the physical and system metrics which are important to understanding photonic interconnection networks. This simplified model enables PhoenixSim simulations to run on conventional computers in a period of minutes or hours. The device characteristics can be determined experimentally, through FDTD simulation, or extrapolated through projection of technological progress. This set of modeled devices composes the *Photonic Device Library*. While the descriptions included in this chapter mostly highlight silicon ring-based topologies, the modeling methodology can easily be used to describe devices based on other technology domains such as Mach-Zehnders, photonic crystals, and micro-electro-mechanical systems (MEMS).

The parameters used to describe basic photonic devices, called *Basic Elements*, are shown in Fig. 4.4. PhoenixSim refers to optical inputs and outputs as ports. Each port is physically bi-directional, therefore ports from which an optical signal can ingress into can also be used to egress from, and vice versa. Certain network topologies may still require uni-directional operation of the ports to facilitate simplicity or satisfy some other design requirement. Nonetheless, the bi-directional nature of each port is still represented for accuracy. The ports of the device are enumerated $0...N-1$ where N is the number of ports of a photonic device. The later figures in this section

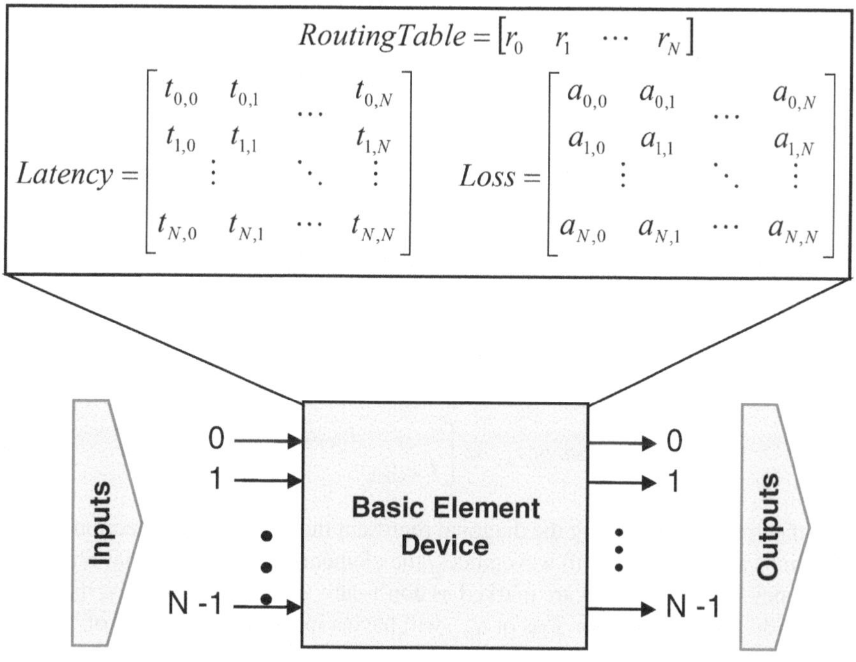

Fig. 4.4 Parameters for characterizing a photonic device using the *Basic Element Model* [1]

which show device geometry will have ports (represented by black dots) labeled with their assigned value. N also determines the size of additional parameter matrices used in defining the photonic device behavior and characteristics.

PhoenixSim uses a logical routing table to determine the path a message takes through the device. Figure 4.4 shows how the routing table can be represented as a length-N vector, where the index represents the ingression port of an optical signal and the value at the index represents the egression port.

Additionally, two tables are used to represent the latency and the optical insertion loss properties of the device. Each property is represented as a $N \times N$ matrix where the row corresponds to the port through which the optical signal ingresses from (input) and the column represents the port from which the optical signal egresses from (output). Each entry in a matrix corresponds to the value used for the particular input/output combination. The latency for a particular input-output port combination is measured as the time between when optical signal enters the input port and when the same optical signal exits the output port. The insertion loss is a measure of the optical power attenuation an optical signal receives when traveling through a device and is useful in characterizing network-level insertion loss and crosstalk.

4.1.3.1 Static Elements

The *Basic Element Model* is most suitable for describing static optical devices that have characteristics that do not change at runtime. The current library of devices focus on 2-D planar devices that are capable of being fabricated in a CMOS-compatible process. These static devices include waveguides, waveguide bends, waveguide crossings, and couplers.

Straight Waveguides. A straight waveguide can be characterized by its length and insertion loss. Propagation loss is affected by a variety of parameters including waveguide dimensions, fabrication technique, and material properties. Waveguides are modeled as 2-port devices with parameters for length, group velocity per unit length, and insertion loss per unit length.

A waveguide's routing table is $[1, 0]$; which indicates that an optical signal ingressing on either end will egress on the opposite side. For a waveguide of length L_{wg} and propagation delay t_{wg}, the latency matrix will be:

$$Latency_{wg} = \begin{bmatrix} - & L_{wg}t_{wg} \\ L_{wg}t_{wg} & - \end{bmatrix} \tag{4.1}$$

Note that the elements along the diagonal represent the latency of a reflection. Since reflections are nonexistent in waveguides, the elements of the matrix that represent the latency of the reflection are marked as don't-care values (–). Similarly, the same waveguide with propagation loss of α_{wg} will have a insertion loss matrix of:

$$Loss_{wg} = \begin{bmatrix} \infty & L_{wg}\alpha_{wg} \\ L_{wg}\alpha_{wg} & \infty \end{bmatrix} \tag{4.2}$$

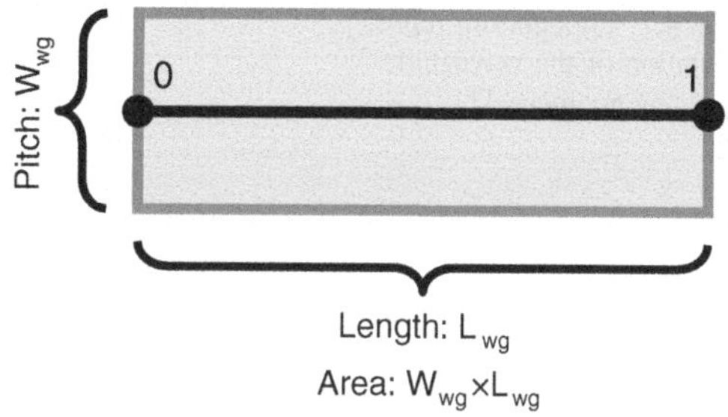

Fig. 4.5 PhoenixSim representation of the straight waveguide geometry [1]

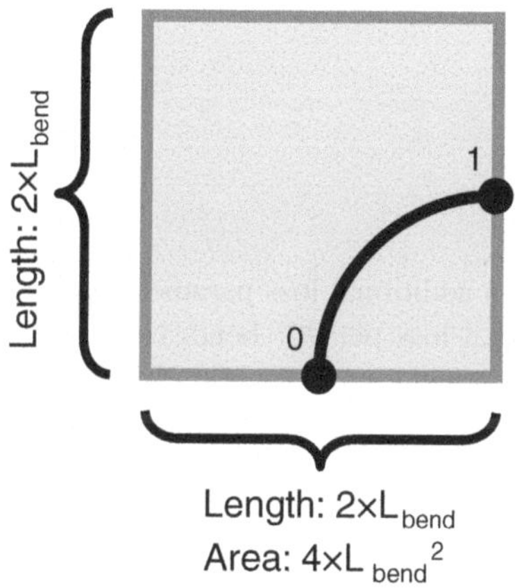

Fig. 4.6 PhoenixSim representation of a 90° bending waveguide geometry [1]

While reflections do not occur in the waveguide, it is useful to assign infinite (∞) insertion loss to the reflection path for crosstalk calculation purposes.

The straight waveguide geometry is shown in Fig. 4.5. Although typical waveguides are less than a micron in pitch, a large buffer spacing needs to be enforced around the waveguide to prevent unintended evanescent coupling and crosstalk. PhoenixSim assumes a default buffer spacing value of 2.5 μm for all devices, therefore the waveguide element utilizes an effective pitch of around 5 μm.

Waveguide Bends. Waveguide bends contribute additional insertion loss to the waveguide's existing propagation loss, which is referred to as bending loss. Bends are modeled as 2-port devices and take parameters for loss per degree and angle of the bend. The bending waveguide element geometry is illustrated in Fig. 4.6.

Similar to straight waveguides, waveguide bends also possess a routing table of [1, 0]. The radius of the waveguide bend, L_{bend}, must be specified. For the purposes of simulation and layout, the PhoenixSim waveguide bend model assumes a bending radius of 2.5 μm and parameterizes the insertion loss according to the angle of the arc, θ_{bend}. The latency matrix for waveguide bends is:

$$Latency_{bend} = \begin{bmatrix} - & \theta_{bend} L_{bend} t_{wg} \\ \theta_{bend} L_{bend} t_{wg} & - \end{bmatrix} \tag{4.3}$$

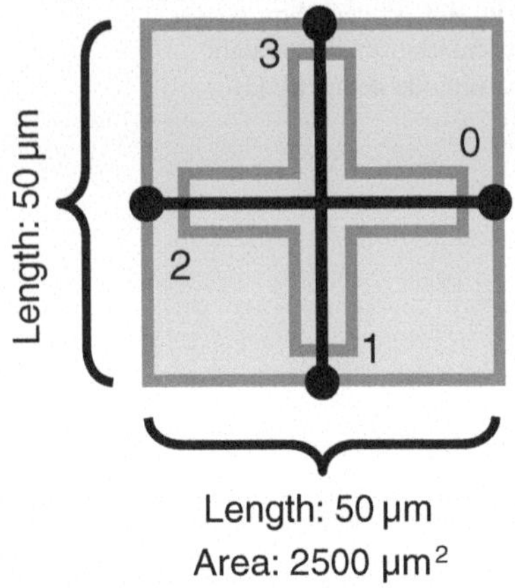

Fig. 4.7 PhoenixSim representation of the waveguide crossing geometry [1]

An additional loss parameter is used for the bending loss, α_{bend}, which defines the total loss per 90° bend. Note that in the PhoenixSim definition, the bending loss parameter includes both the propagation loss of the waveguide as well as the excess loss due to the bend. This produces a loss matrix for bends as follows:

$$Loss_{bend} = \begin{bmatrix} \infty & 2\alpha_{bend}\theta_{bend}/\pi \\ 2\alpha_{bend}\theta_{bend}/\pi & \infty \end{bmatrix} \tag{4.4}$$

In the same fashion as the straight waveguide, the design building block of the bending waveguide requires an area much larger than the waveguide itself to prevent coupling.

Waveguide Crossings. The model for crossings are configured as 4-port devices with parameters for the loss and crosstalk. Unlike the straight and bending waveguides previously described, the waveguide crossing is the first element thus far described that exhibits crosstalk, which is the act of inducing noise on a signal. The routing table is [2, 3, 0, 1]. The ordering of the indexes of the routing table are labeled in Fig. 4.7 and correspond to the cardinal directions in the following order: East, South, West, and North.

The PhoenixSim waveguide crossing model assumes the design described by Bogaerts et al. [9]. The crossing is double etched at the intersection to create a mode expanding region to reduce the loss, crosstalk, and back reflection. PhoenixSim assumes a fixed crossing length of 50 μm for each waveguide, crossing at exactly the midpoint.

The latency matrix of the waveguide crossing is as follows:

$$Latency_{cross} = \begin{bmatrix} 50\,\mu\text{m}\cdot t_{wg} & 50\,\mu\text{m}\cdot t_{wg} & 50\,\mu\text{m}\cdot t_{wg} & 50\,\mu\text{m}\cdot t_{wg} \\ 50\,\mu\text{m}\cdot t_{wg} & 50\,\mu\text{m}\cdot t_{wg} & 50\,\mu\text{m}\cdot t_{wg} & 50\,\mu\text{m}\cdot t_{wg} \\ 50\,\mu\text{m}\cdot t_{wg} & 50\,\mu\text{m}\cdot t_{wg} & 50\,\mu\text{m}\cdot t_{wg} & 50\,\mu\text{m}\cdot t_{wg} \\ 50\,\mu\text{m}\cdot t_{wg} & 50\,\mu\text{m}\cdot t_{wg} & 50\,\mu\text{m}\cdot t_{wg} & 50\,\mu\text{m}\cdot t_{wg} \end{bmatrix} \tag{4.5}$$

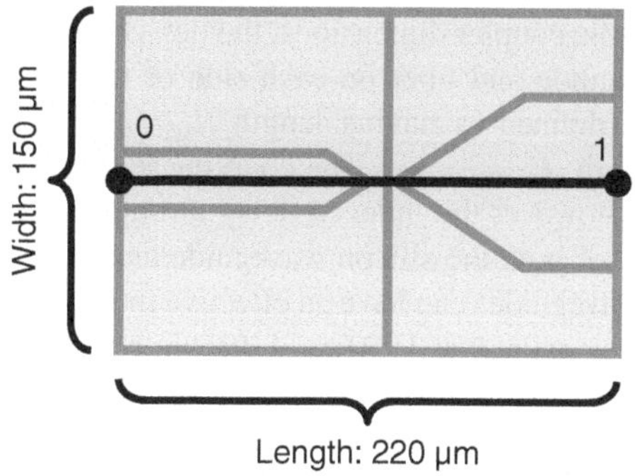

Fig. 4.8 PhoenixSim representation of an example coupler geometry, connecting a silicon waveguide to a tapered fiber. The width along the lateral direction of the interface is dominated by the fiber diameter. The length accounts for the tapering at the fiber tip (*left*) and the inverse taper of the waveguide in the silicon substrate (*right*) [1]

Observe that the latency takes the value $50\,\mu\text{m} \cdot t_{wg}$ for all possible input-output path combinations. Due to the symmetry that exists along each of the four arms of the crossing, any optical signal will always propagate along two arm segments (25 μm long, each).

The PhoenixSim model of waveguide crossings is based also on the following parameters: insertion loss, α_{cross}, crosstalk, α'_{cross}, and back reflection, R_{cross}. The loss matrix for the waveguide crossing is:

$$Loss_{cross} = \begin{bmatrix} R_{cross} & \alpha'_{cross} & \alpha_{cross} & \alpha'_{cross} \\ \alpha'_{cross} & R_{cross} & \alpha'_{cross} & \alpha_{cross} \\ \alpha_{cross} & \alpha'_{cross} & R_{cross} & \alpha'_{cross} \\ \alpha'_{cross} & \alpha_{cross} & \alpha'_{cross} & R_{cross} \end{bmatrix} \tag{4.6}$$

Only the logically correct paths (north-to-south, south-to-north, east-to-west, and west-to-east) take the insertion loss value. Paths that must 'turn' at the intersection observe the crosstalk loss. Reflections (matrix diagonal) take on a non-negligible value unlike in the previous waveguide examples.

Couplers. Optical couplers are modeled as a 2-port device with a single parameter for insertion loss. Figure 4.8 shows an example coupling interface between an on-chip silicon waveguide and an off-chip single-mode silica fiber. The routing table is $[1, 0]$.

The insertion loss of the coupler, $\alpha_{coupler}$, predominantly comes from the scattering and reflection, $R_{coupler}$. The insertion loss matrix for a coupler is defined as:

$$Loss_{coupler} = \begin{bmatrix} R_{coupler} & \alpha_{coupler} \\ \alpha_{coupler} & R_{coupler} \end{bmatrix} \tag{4.7}$$

When considering the coupler dimensions, the device might require special conditioning of the waveguide and fiber on each side of the interface. Therefore the waveguide portion is defined as having length, $L_{coupler.wg}$, and the fiber side is defined as having length, $L_{coupler.fiber}$. The coupling length is a summation of two waveguide segments, however the induced delay through the coupler is dependent on the effective index of both the silicon waveguide and the silica fiber. The high-confinement silicon waveguides can have an effective index of over 4 (highly dependent on waveguide dimensions) at 1550 nm [10] while SSMF possess an effective index of 1.47 at 1550 nm [11]. This large index contrast requires the assumption of a different propagation delay for the fiber side, t_{fiber}.

The coupler latency matrix is defined as follows:

$$Latency_{coupler} = \begin{bmatrix} 2t_{wg}L_{coupler.wg} & t_{fiber}L_{coupler.fiber} + t_{wg}L_{coupler.wg} \\ t_{fiber}L_{coupler.fiber} + t_{wg}L_{coupler.wg} & 2t_{fiber}L_{coupler.fiber} \end{bmatrix} \quad (4.8)$$

4.1.3.2 Ring-Resonator Elements

As described in Chap. 3, ring resonators are extremely versatile structures that can be used to implement many network functions. To model the various ring resonator devices, PhoenixSim extends the *Basic Element Model* with subclasses for *Ring Elements* and *Dynamic Elements* (Fig. 4.9). The *Dynamic Element Model* is used to describe active devices which can exhibit changes in its routing table, latency matrix, and loss matrix during runtime. The properties of the active device during its operation is defined by *state* variables which can be changed and controlled. The *Ring Element Model* supports the definition of the resonant behavior of the devices. The behavior of ring-based devices is determined by the wavelength of the optical signal that interacts with the component. Also shown in Fig. 4.9 is how *Dynamic-Ring Elements* can be derived from the individual *Ring* and *Dynamic Element*. For instance, a ring-based broadband switch consists of a combination of ring resonators and electrical logic (described later) and can be electro-optically controlled to alter the optical flow of data.

Filters. Optical filters are useful in selectively extracting a subset of wavelengths from a WDM message. In the limiting case, an extremely small ring will have a large FSR and allow the filtering of a single wavelength channel. Filtering is accomplished by aligning the spectral mode of the ring with the wavelength channel of interest (Fig. 4.10a). Light at wavelengths that align with the mode of the ring (on resonance) will couple from the ingression waveguide, into the ring structure, and out onto a secondary waveguide; wavelengths of light that are not aligned (off resonance) will be unperturbed by the ring and continue down the injection waveguide (Fig. 4.10c). PhoenixSim models ring filter devices as single-state 4-port *Ring Elements* with a parameter for the ring diameter (assuming a circle). Ring filters have been fabricated and demonstrated on SOI with 3-μm radius, corresponding to an FSR of 30 nm [12].

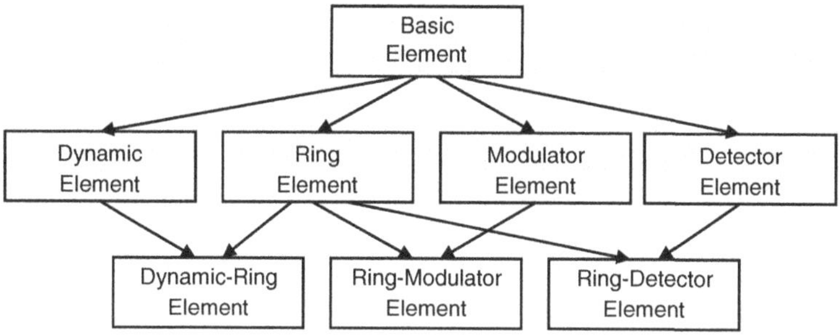

Fig. 4.9 Organization of building block element classes within PhoenixSim [1]

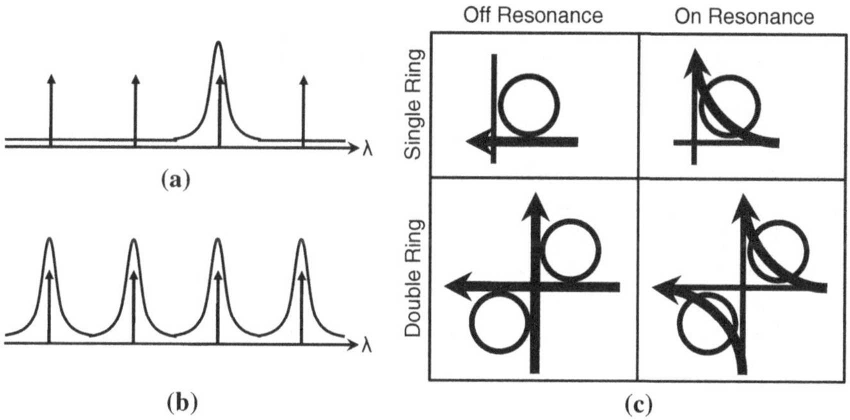

Fig. 4.10 Propagation through a ring-resonator device depends on the signal wavelength and the resonant modes of the device. **a** Small rings with larger mode spacings (shown as periodic peaks) can be designed to interact with a single wavelength channel from a WDM signal (indicated by *arrows*). **b** Broadband switch have tightly spaced modes, enabling many WDM channels to couple into the device cohesively. **c** The path of propagation depends on whether the wavelength of the message is on- or off-resonance with the ring [1]

Broadband Switches. Ring resonators are also capable of controlling the flow of an entire WDM message by aligning each wavelength channel to a mode of the ring (Fig. 4.10b). This can be accomplished in a limited spectral range by using a large ring with a correspondingly small FSR. When all the wavelength channels are on resonance, the entire WDM message will couple into the ring and onto a second waveguide, similar to the case of the filter. Additionally, if the FSR is manipulated electro-optically, all the modes can be shifted so that the wavelength channels are no longer on resonance, thus causing the entire WDM message to not couple into the ring. This functionality is illustrated in Fig. 4.10c for both a single-ring 1×2 photonic switching element (PSE) and a double-ring 2×2 PSE. These broadband switch elements are modeled as two-state 4-port devices. A 1×2 switch composed of a ring with a 100-μm radius and 0.8-nm FSR was shown to be capable of switching

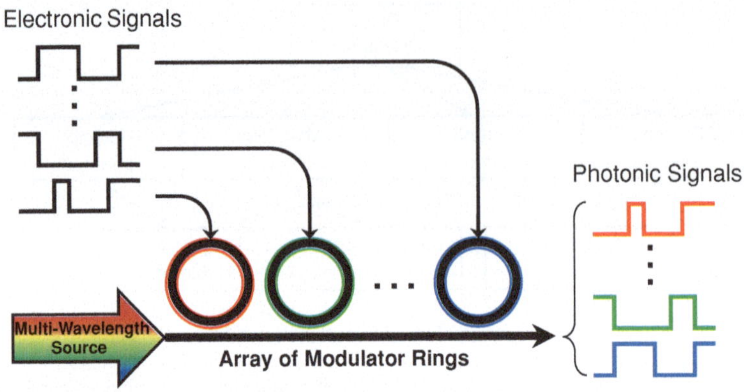

Fig. 4.11 Schematic of the conversion process between the spatially-parallel electronic domain and wavelength-parallel optical domain [1]

20 wavelength channels simultaneously [13]. Elsewhere, a fifth-order switch was demonstrated being able to simultaneously route nine 40-Gbps wavelength channels for an aggregate data rate of 360 Gbps [14].

Modulators. Ring-based modulators are essentially high-speed switches. By electro-optically flipping the ring between an on- and off-resonance state, a series of 0's and 1's can be encoded onto an optical stream of light. Light that couples into the ring will not egress into another waveguide like the filters and switches, but will eventually dissipate within the ring. A modulator array can be formed with multiple ring modulators so that several wavelength channels can be encoded in parallel, creating a WDM signal (Fig. 4.11). Modulators should have a small ring diameter to create a large FSR to ensure that the modulation does not interfere with other spectrally adjacent wavelength channels. The modulator device is modeled as a single-state device with parameters for energy dissipated per modulated bit and ring diameter. Ring-based modulation has been demonstrated at rates of 12.5 Gbps in a 5-μm radius silicon ring resonator [15].

Receivers (Photo-Detectors). PhoenixSim *DetectorElements* assume that a ring filter is placed before the photo-detector element for selecting specific wavelengths from a WDM signal. The *detector sensitivity* determines the minimum signal power that must be received at the photo-detector in order for data to be properly recovered from the optical domain and is an important parameter for determining the optical power budget (as discussed in Sect. 4.1.4). This ring-based detection device takes parameters for energy dissipated per detected bit, sensitivity, and ring diameter. Integrated high-speed germanium detectors have been demonstrated operating at speeds of 40 Gbps [16, 17].

4.1.3.3 Mach-Zehnder Elements

Switches and modulators can also be designed using the principle of Mach-Zehnder interferometry (MZI). Mach-Zehnder devices are designed to operate relatively uniformly over a large wavelength range and do not exhibit the sharp resonant peaks that ring resonators have. For instance, a MZI-based device can be used to modulate wavelengths of light that span a large continuous wavelength range while ring-resonator modulators are limited to specific resonance wavelengths. However, this operational difference between Mach-Zehnder devices and ring-resonator devices causes them to not be interchangeable. Models for 1×2 and 2×2 Mach-Zehnder switches are currently included in the *Photonic Device Library*. A modulator and switch based on MZI has been demonstrated operating at up to 10 Gbps [18].

4.1.4 Physical-Layer Performance Analysis Tools

The consideration of the photonic technology domain presents new design challenges that must be satisfied in order to produce feasible interconnect designs. Similar to electronics, it is important for photonic networks to consider power dissipation and system-level performance. Furthermore, photonic networks must also consider metrics that have no electronic equivalent such as insertion loss, the optical power budget, noise, and crosstalk. While a comprehensive analysis of a photonic interconnect design would involve the actual fabrication and operation of such a system, this is currently unrealistic since full-scale photonic on-chip networks are still in early stages of research. Therefore, the tools presented here can give important insight into the physical feasibility of the designs and the performance that is expected.

4.1.4.1 Optical Power Budget

The optical power budget of a photonic network assesses the amount of WDM parallelism and insertion loss that can be tolerated. Many currently proposed photonic interconnection networks assume off-chip lasers to provide the optical sources, which are then coupled into the chip where they are modulated, routed, and received. Optical amplification in an on-chip environment is not easily accomplished in the CMOS platform. For this reason, the power that is received at the photo-detectors must remain above a certain power threshold (labeled the detector sensitivity in Fig. 4.12) to ensure proper detection of data bit streams. This limitation can be partially compensated by increasing the optical power that is injected into the chip. However, this also exhibits an upper limitation due to nonlinearities of the silicon material which could potentially distort the signal. Distortions are caused by nonlinearities within silicon which contribute additional insertion losses and can also cause unwanted shifts in the resonances of ring resonators. This limit is labeled as *nonlinear effects* in Fig. 4.12. The difference in the two thresholds is called the *optical power budget*.

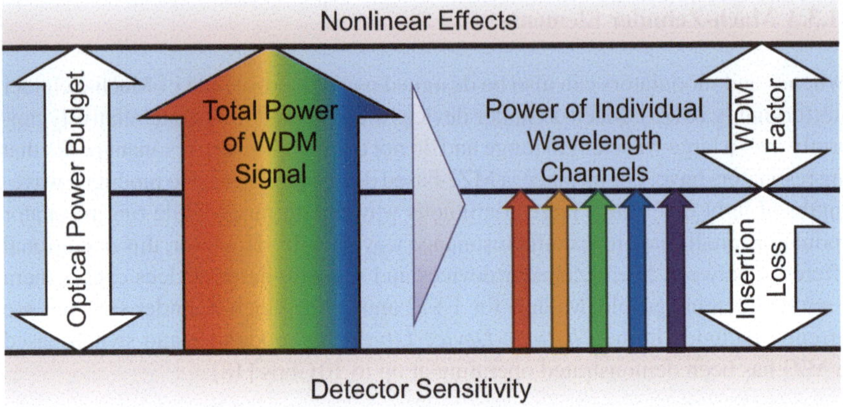

Fig. 4.12 The relationship of various parameters affecting the optical power budget. The difference in power of the total WDM signal (*large arrow* on the *left*) and the individual wavelength channels (*five smaller arrows* on the *right*) constrains the scalability of the system [1]

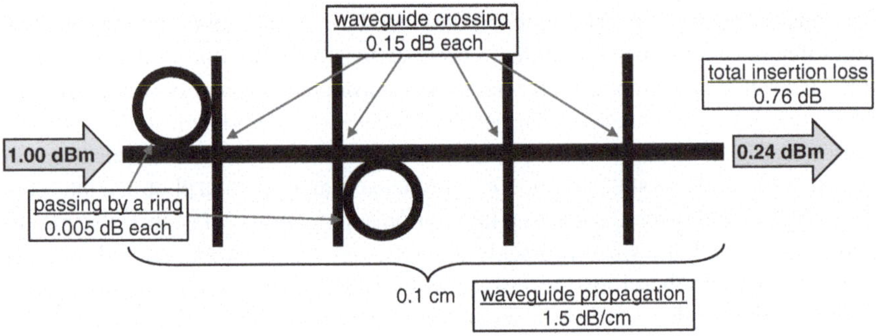

Fig. 4.13 Calculation of insertion loss for a small network segment [1]

As shown in Fig. 4.12, the optical power budget affects the design choices of a given network architecture by constraining the sum of the WDM factor and the network insertion loss. The *WDM factor* measures the power difference between an entire WDM signal and its constituent wavelength channels. This factor needs to be evaluated since the nonlinearity threshold is determined by the total power in the waveguide while the detector sensitivity depends on the power in the individual wavelengths. The remaining portion of the optical power budget must accommodate the worst-case insertion loss that an optical message could receive in the network. Figure 4.13 shows an example of the calculation involved in determining the insertion loss for an optical signal being injected into a small network segment at 1 dBm. The signal is ejected at 0.24 dBm after propagating across a 0.1-cm distance, passing by two ring resonators, and entering four waveguide crossings. The total loss for this example is 0.76 dB. For a full-scale photonic network, all valid optical paths need to be examined to determine the path with the highest loss.

The relationship between the various device limitations and system-level metrics is summarized in the inequality

$$P - S \geq IL_{max} + 10log_{10}n \tag{4.9}$$

where P is the limit in optical power of the system and S is the detector sensitivity. P represents the maximum optical power that can be inserted into the photonic system without inducing adverse behavior in the photonic devices. S represents the minimum amount of power that can be received by a detector while maintaining certain error tolerance. The optical power budget is $P - S$. The worst-case optical path in terms of insertion loss is IL_{max} and n specifies the number of wavelength channels being used (WDM factor). P, S, and IL_{max} are expressed in decibel units.

While it may be desirable to maximize the number of wavelength channels used to increase bandwidth through parallelism, and to create scalable photonic networks at the cost of higher insertion losses, Eq. 4.9 shows the inherent limitations of this approach. From an architectural standpoint, P and S are fundamental design constraints imposed by the photonic devices. Therefore, a designer must strike a balance between the desired link bandwidth and the desired complexity of the network.

4.1.4.2 Data Integrity

A variety of interactions in a photonic interconnection network will work to degrade the integrity of transmitted data. The PhoenixSim noise modeling methodology accounts for intensity noise generated at the laser sources, inter-message crosstalk, intra-message crosstalk, and electrical noise generated by the optical receivers (Fig. 4.14). The standard figure of merit for measuring the quality of signal is the signal-to-noise ratio (SNR) which is defined as the ratio between signal power and noise power. More specifically, the optical SNR (OSNR) is the ratio of optical signal power to optical noise power at the point where the measurement is being made. From a system perspective, the SNR can be used to determine the statistical likelihood that each bit of data is transmitted erroneously (e.g. a transmitted 0 is detected as a 1), also called *bit error rate* (BER). An understanding of the potential noise in any interconnection network is critical to determining the effective throughput of the system since error detection and correction will invariably cause performance overheads.

The first source of noise is the laser sources which inherently causes random fluctuations in an optical signal, called intensity noise. This noise is quantified as relative intensity noise (RIN), which is the ratio of the power variance of the optical signal to the mean optical power squared. Quantum cascade lasers have a measured RIN on the order of −150 dB/Hz with an output of 10-dBm mean optical power [19]. To convert to an SNR, the following relation is used [20]:

$$SNR_{laser} = \frac{m^2}{2\,\mathrm{B} \cdot RIN} \tag{4.10}$$

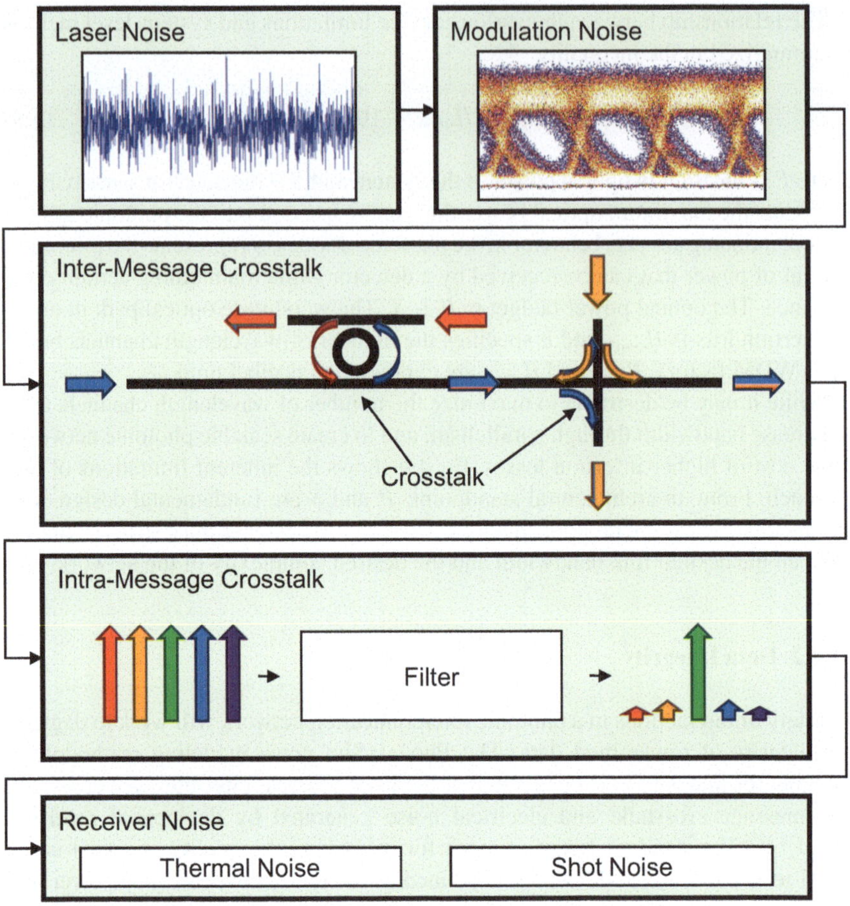

Fig. 4.14 Sources of noise and crosstalk within a chip-scale photonic system [1]

where B is the noise bandwidth, assumed equal to the modulation rate, and m is the modulation index, equal to $1 - E$, where E is the extinction ratio of the modulator.

A second source of noise is *inter-message crosstalk* which occurs when multiple photonic messages concurrently propagate through a photonic device. In a waveguide crossing for example, the ideal situation is for two orthogonally propagating messages to be completely isolated from each other with no interaction. However, in reality a small amount of optical power from each message will leak onto the other message. A similar situation occurs in ring-resonator filters and switches due to imperfect coupling of each wavelength channel.

For the N-port device, the crosstalk power that a message on a particular port receives is given by the sum of the power that is leaked by any existing messages on the other $N-1$ ports. If M is the set of all signals present in the device and the power of a signal k is given by the variable P_k, then the crosstalk power seen by signal s is given by

$$\sum_{k \in M, k \neq s} \frac{P_k}{IL(portin_k, portout_s)} \tag{4.11}$$

which aggregates the unwanted signal power that leaks into the output port being used by s. Function IL refers to the *insertion-loss matrix* (that was described in Sect. 4.1.3) of the device model with arguments for the input and output port. In Eq. 4.11, $portin_k$ denotes the input port of a message k, and $portout_s$ denotes the output port of s. This calculation is a first-order approximation that only considers crosstalk for messages that coexist in a device and not from leaked power that propagates across multiple devices before interfering with a foreign signal.

A third source of noise called *intra-message crosstalk* occurs due to imperfect filtering. For example, in order for a WDM message to be received and converted into an electrical signal, each wavelength channel must be individually filtered and fed into a photo-detector. Due to imperfect extinction, power from the adjacent wavelength channels will leak through, causing an additional source of noise. Intra-message crosstalk will also occur in any other location in a photonic network where filtering functionality is involved. The spectral response of a ring resonator mimics a periodic Lorentzian function. For simplicity, PhoenixSim assumes a periodic flat passband and constant extinction ratio for the stop bands. Lastly, the receiver model includes thermal noise and shot noise.

The combined effect of these multiple sources of noise can be used to compute an SNR for the final detected signal with the following equation:

$$SNR = \frac{P}{N_{laser} + N_{inter} + N_{intra} + N_{therm} + N_{shot}} \tag{4.12}$$

where P is the signal power and N corresponds to the noise power associated with the noise or crosstalk source indicated by the subscript.

4.1.4.3 Power Dissipation

To compute the power dissipation of the modeled networks, PhoenixSim takes the sum of the energy dissipation events from all devices. The photonic device library tracks the power dissipation according to the type of model that is used, and can include both static (over a duration of time) and dynamic (instantaneous) power dissipation. *Dynamic Element* devices can have static power dissipation, which is determined by the occupied state. *Dynamic Element* devices can also have dynamic power dissipation, which is accumulated whenever there is a state transition. An additional source of power dissipation are *Ring Element* devices, which require constant thermal tuning to compensate for fabrication uncertainty and ambient temperature shifts. *Modulator* and *Detector Elements* also dissipate power during the transmission and detection of data, respectively.

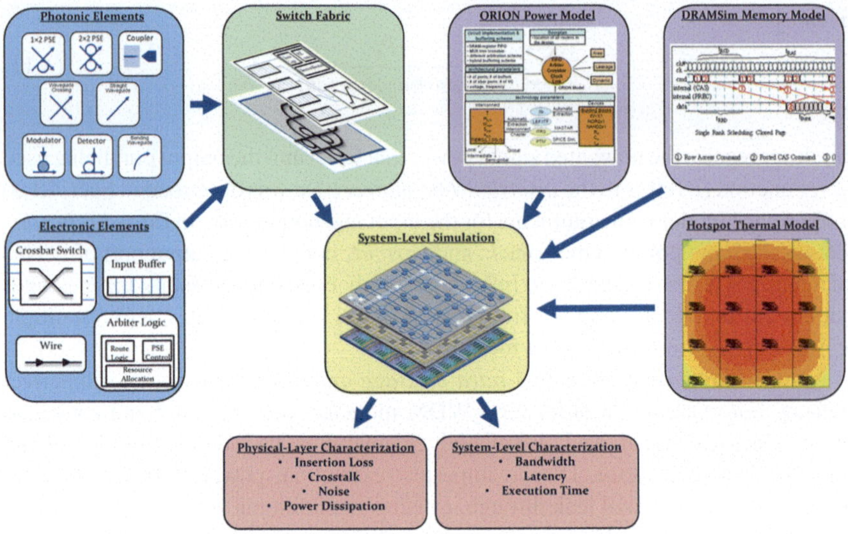

Fig. 4.15 Organization of the PhoenixSim environment

Electronic routers are modeled as standard three-stage pipelines. The power modeling of the electronic routers is accomplished by leveraging the ORION simulator, which is currently capable of modeling down to the 32-nm technology node [3].

4.1.5 Integration with Other Simulators

In addition to the PhoenixSim code base, a number of third-party tools and simulators are also integrated and leveraged. This enables a user of PhoenixSim to model and simulate networks with a richer and more detail level of precision. Figure 4.15 shows the organization of PhoenixSim with these integrated third-party tools. As mentioned before, PhoenixSim integrates ORION for its electronic router power model [3]. For modeling memory, the simulator has been interfaced with DRAMSim which was developed at University of Maryland [21]. PhoenixSim can also operate together with the Hotspot thermal simulator developed at University of Virginia [22, 23].

4.2 Related Work

As the interest for using photonic interconnects continues to grow, so does the need for tools that can harness the potential of this new technology. In the realm of simulation, two levels exist which are of interest to photonic network designers: link-level and system-level. Simulation is an especially important predictive tool for gauging the performance of these photonic interconnect systems which are too complex for manufacturing in current fabrication technology. Beyond simulation, design tools will be needed to effectively and accurately design complex and efficient photonic interconnection networks. Most conventional simulation and design tools are not ideally suited for capturing the physical and performance characteristics of chip-scale photonic interconnection devices and networks. Therefore the development of photonically-enabled tools is needed to fill the void.

As photonic interconnect topologies are becoming increasingly complex, layout tools and optimization techniques will be required for efficient and accurate design. Ding et al. have developed OIL (Optical Interconnect Library) a synthesis-like CAD tool for optimizing optical router designs in terms of insertion loss [24]. The methodology allows for constraint-based optimization in terms of latency and insertion loss. Similarly, Minz et al. have devised a synthesis tool for timing-driven optimization of optical waveguide placement in an on-chip network [25]. VANDAL is a place-and-route tool for on-chip photonic architectures which uses a library of modeled and characterized components, and includes automation tools for rapid design and synthesis [26].

With link-level simulation, the primary concern is detailed physical modeling of all the end-to-end aspects of a photonic path to determine performance metrics such as signal integrity and link reliability. O'Connor et al. proposed a link-level simulation environment for heterogeneous photonic integrated circuits which leverages detailed synthesizable models of building-block components for the purpose of determining interconnect density, area, link delay, and link power requirements [27]. Similarly, De Wilde et al. presented an approach for characterizing CMOS-to-CMOS links in terms of timing, error rates, and noise sensitivity [28]. The IBM optical link simulator was created to design and analyze telecom- and LAN-scale links through metrics such as failure rates, power penalties, and signal performance (e.g. eye diagrams) [29].

System-level simulation uses a higher-level of abstraction than link-level simulation and is primarily concerned with determining network performance metrics (e.g. bandwidth, application latency, and system power dissipation). Briere et al. have developed the ONoC SystemC model which focuses on the simulation of optical networks-on-chip using the SystemC framework and primarily addressing high-level system concerns including device timing and network-level power dissipation [30]. Their modeling is currently specific to topologies that leverage the *lambda router*, which routes optical traffic based on the wavelength of light that is used by the source. OptiSim is a system-level simulator for modeling optical interconnects in board- and cluster-based computing [31].

The PhoenixSim environment closely resembles OptiSim (Ref. [31]) with respect to the use of a photonic building block library, and extractability of physical and system metrics. PhoenixSim differs from OptiSim through a combination of the focus on chip-scale systems, support for spatial and temporal based photonic chip-scale architectures, and synergistic study of physical-layer and system-level performance metrics.

References

1. J. Chan, G. Hendry, K. Bergman, and L. Carloni, "Physical-layer modeling and system-level design of chip-scale photonic interconnection networks," *Computer-Aided Design of Integrated Circuits and Systems, IEEE Transactions on*, vol. 30, no. 10, pp. 1507–1520, Oct. 2011.
2. A. Varga, "OMNeT++ discrete event simulation system." [Online]. Available: http://www.omnetpp.org.
3. H.-S. Wang, X. Zhu, L.-S. Peh, and S. Malik, "Orion: a power-performance simulator for interconnection networks," in *Proceedings of the 35th Annual ACM/IEEE International Symposium on Microarchitecture (MICRO)*, Nov. 2002, pp. 294–305.
4. D. Vantrease, R. Schreiber, M. Monchiero, M. McLaren, N. P. Jouppi, M. Fiorentino, A. Davis, N. Binkert, R. G. Beausoleil, and J. H. Ahn, "Corona: System implications of emerging nanophotonic technology," *Proceedings of the 35th Annual International Symposium on Computer Architecture (ISCA)*, pp. 153–164, Jun. 2008.
5. A. Shacham, K. Bergman, and L. P. Carloni, "Photonic networks-on-chip for future generations of chip multiprocessors," *IEEE Transactions on Computers*, vol. 57, no. 9, pp. 1246–1260, 2008.
6. G. Hendry, S. Kamil, A. Biberman, J. Chan, B. Lee, M. Mohiyuddin, A. Jain, K. Bergman, L. Carloni, J. Kubiatowicz, L. Oliker, and J. Shalf, "Analysis of photonic networks for a chip multiprocessor using scientific applications," in *Proceedings of the 2009 3rd ACM/IEEE International Symposium on Networks-on-Chip (NOCS)*, May 2009, pp. 104–113.
7. P. Dong, S. F. Preble, and M. Lipson, "All-optical compact silicon comb switch," *Opt. Express*, vol. 15, no. 15, pp. 9600–9605, Jul. 2007.
8. J. Chan, G. Hendry, A. Biberman, and K. Bergman, "Architectural exploration of chip-scale photonic interconnection network designs using physical-layer analysis," *J. Lightwave Technol.*, vol. 28, no. 9, pp. 1305–1315, May 2010.
9. W. Bogaerts, P. Dumon, D. V. Thourhout, and R. Baets, "Low-loss, low-cross-talk crossings for silicon-on-insulator nanophotonic waveguides," *OSA Optics Letters*, vol. 32, no. 19, pp. 2801–2803, 2007.
10. A. Sakai, G. Hara, and T. Baba, "Large effective index and low bend loss in SOI optical waveguides," in *Lasers and Electro-Optics, 2001. CLEO/Pacific Rim 2001. The 4th Pacific Rim Conference on*, vol. 1, 2001, pp. I-4–I-5.
11. Corning Incorporated, "Corning SMF-28e+ optical fiber product information," Jul. 2011. [Online]. Available: http://www.corning.com/WorkArea/showcontent.aspx?id=41261.
12. B. Little, J. Foresi, G. Steinmeyer, E. Thoen, S. Chu, H. Haus, E. Ippen, L. Kimerling, and W. Greene, "Ultra-compact Si-SiO_2 microring resonator optical channel dropping filters," *IEEE Photonics Technology Letters*, vol. 10, no. 4, pp. 549–551, Apr. 1998.
13. B. G. Lee et al., "All-optical comb switch for multiwavelength message routing in silicon photonic networks," *IEEE Photonics Technology Letters*, vol. 20, no. 10, pp. 767–769, May 2008.
14. Y. Vlasov, W. M. J. Green, and F. Xia, "High-throughput silicon nanophotonic wavelength-insensitive switch for on-chip optical networks," *Nature Photonics*, vol. 2, pp. 242–246, Apr. 2008.
15. Q. Xu, S. Manipatruni, B. Schmidt, J. Shakya, and M. Lipson, "12.5 Gbit/s carrier-injection-based silicon micro-ring silicon modulators," *OSA Optics Express*, vol. 15, no. 2, pp. 430–436, 2007.
16. S. Assefa, F. Xia, S. W. Bedell, Y. Zhang, T. Topuria, P. M. Rice, and Y. A. Vlasov, "CMOS-integrated 40 GHz germanium waveguide photodetector for on-chip optical interconnects," in *Proceedings of Optical Fiber Communication Conference (OFC)*, 2009, p. OMR4.
17. L. Vivien, J. Osmond, J.-M. Fédéli, D. Marris-Morini, P. Crozat, J.-F. Damlencourt, E. Cassan, Y. Lecunff, and S. Laval, "42 GHz P.I.N germanium photodetector integrated in a silicon-on-insulator waveguide," *OSA Optics Express*, vol. 17, no. 8, pp. 6252–6257, 2009.
18. H.-W. Chen, Y.-H. Kuo, and J. E. Bowers, "High speed hybrid silicon evanescent Mach-Zehnder modulator and switch," *OSA Optics Express*, vol. 16, no. 25, pp. 20 571–20 576, 2008.

19. T. Gensty, W. Elsäßer, and C. Mann, "Intensity noise properties of quantum cascade lasers," *OSA Optics Express*, vol. 13, no. 6, pp. 2032–2039, 2005.
20. C. Miller, *Fiber Optic Test and Measurement*. Prentice Hall, 1998.
21. D. Wang, B. Ganesh, N. Tuaycharoen, K. Baynes, A. Jaleel, and B. Jacob, "DRAMsim: a memory system simulator," *SIGARCH Comput. Archit. News*, vol. 33, no. 4, pp. 100–107, Nov. 2005.
22. W. Huang, S. Ghosh, S. Velusamy, K. Sankaranarayanan, K. Skadron, and M. Stan, "Hotspot: a compact thermal modeling methodology for early-stage VLSI design," *Very Large Scale Integration (VLSI) Systems, IEEE Transactions on*, vol. 14, no. 5, pp. 501–513, May 2006.
23. K. Skadron, M. R. Stan, K. Sankaranarayanan, W. Huang, S. Velusamy, and D. Tarjan, "Temperature-aware microarchitecture: Modeling and implementation," *ACM Trans. Archit. Code Optim.*, vol. 1, no. 1, pp. 94–125, Mar. 2004.
24. D. Ding and D. Z. Pan, "OIL: a nano-photonics optical interconnect library for a new photonic networks-on-chip architecture," in *Proceedings of the 11th International Workshop on System Level Interconnect Prediction (SLIP)*, Jul. 2009, pp. 11–18.
25. J. Minz, S. Thyagaraja, and S. K. Lim, "Optical routing for 3D system-on-package," in *Design, Automation Test in Europe Conference Exhibition (DATE)*, vol. 1, Mar. 2006, pp. 1–2.
26. G. Hendry, J. Chan, L. P. Carloni, and K. Bergman, "VANDAL: A tool for the design specification of nanophotonic networks," in *Design, Automation Test in Europe Conference Exhibition (DATE)*, Mar. 2011.
27. I. O'Connor, F. Tissafi-Drissi, F. Gaffiot, J. Dambre, M. De Wilde, J. Van Campenhout, D. Van Thourhout, J. Van Campenhout, and D. Stroobandt, "Systematic simulation-based predictive synthesis of integrated optical interconnect," *IEEE Trans. Very Large Scale Integr. Syst.*, vol. 15, pp. 927–940, Aug. 2007.
28. M. De Wilde, O. Rits, W. Meeus, H. Lambrecht, and J. Van Campenhout, "Integration of modeling tools for parallel optical interconnects in a standard EDA design environment," in *Design, Automation Test in Europe Conference Exhibition (DATE)*, Feb. 2004.
29. P. K. Pepeljugoski and D. M. Kuchta, "Design of optical communications data links," *IBM Journal of Research and Development*, vol. 47, no. 2.3, pp. 223–237, Mar. 2003.
30. M. Briere, E. Drouard, F. Mieyeville, D. Navarro, I. O'Connor, and F. Gaffiot, "Heterogeneous modelling of an optical network-on-chip with SystemC," in *Proceedings of the 16th IEEE International Workshop on Rapid System Prototyping (RSP)*, Jun. 2005, pp. 10–16.
31. A. Kodi and A. Louri, "Optisim: A system simulation methodology for optically interconnected HPC systems," *IEEE Micro*, vol. 28, no. 5, pp. 22–36, Sep.-Oct. 2008.

Chapter 5
Photonic Network Architectures I: Circuit Switching

5.1 Photonic Network Architecture Overview

A photonic network design encompasses a wide range of details that must be considered carefully. Design details include material system, layout of components, and network arbitration mechanisms. Each detail involves several metrics including performance requirements, power constraints, scalability, and cost. Designing a photonic network which is both feasible and cost-effective is a multi-dimensional design-space problem in which all the parts are tightly coupled.

One way to cope with design complexity is through abstraction. Electronic CMOS design has a long history of effectively using layers of abstraction to cope with design complexity. Layers of abstraction range from low-level standard cell libraries to hardware description languages (e.g. VeriLog, VHDL) to system level modeling (e.g. SystemC). Similarly PhoenixSim and its Interconnect Building Block Library (Sect. 4.1.3) provides a similar form of abstraction to facilitate easier network design. However, a key difference between CMOS design and current photonic network design is that there is a strong relationship between the physical layer characteristics and system performance. Changing the characteristics of a single photonic device can induce performance and layout throughout the entirety of the system. For this reason, photonic network design requires an all-encompassing design methodology, taking into account everything from device behavior to networking concepts to the application running on the system.

The high-level structure of many chip-scale photonic network architectures that have been proposed in literature is illustrated in Fig. 5.1. The chip is composed of three functional layers: a processing plane, an electronic control plane, and a photonic data plane. The processing plane is where the processing nodes sit and act as the sources and sinks for all communications. The top most layer, the photonic data plane, provides optical links between any pair of communicating processors. However, because the photonic plane cannot easily perform logical processing, an electronic control plane is provided for the purposes of configuration and routing.

K. Bergman et al., *Photonic Network-on-Chip Design*,
Integrated Circuits and Systems, DOI: 10.1007/978-1-4419-9335-9_5,

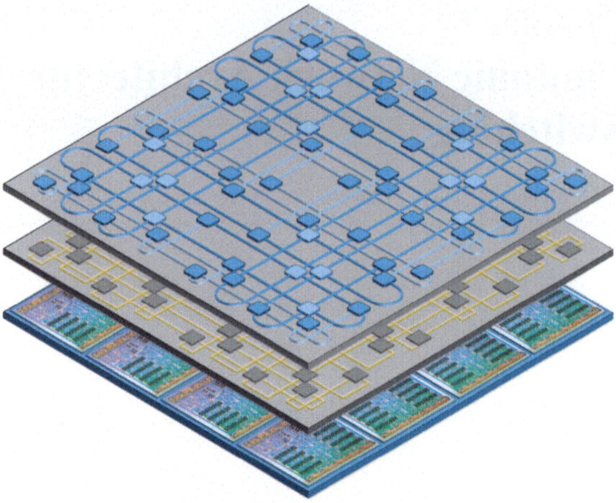

Fig. 5.1 The envisioned chip stack for chip-scale photonic interconnection networks. Three primary layers consisting of a processing layer (*bottom*), electronic control plane layer (*middle*), and photonic data plane layer (*top*) [1]

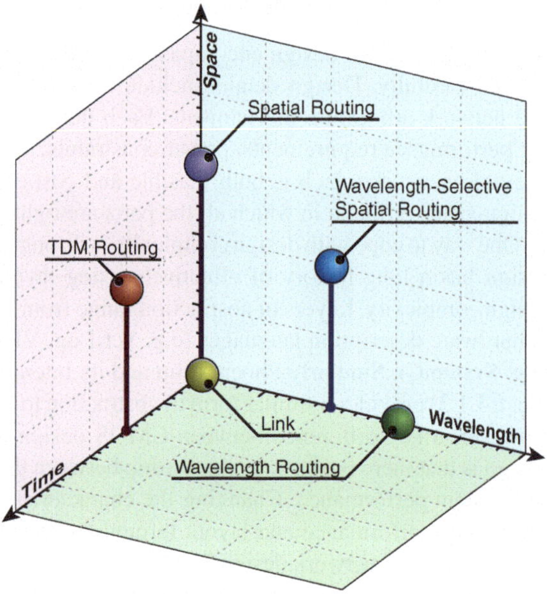

Fig. 5.2 Routing technique design space based on three arbitration domains: time, wavelength, and space

The various proposed photonic networks can be generally classified as leveraging a combination of three optical routing domains: time, wavelength, and space. Figure 5.2 illustrates the design space that is afforded by these three domains and the relative placement of the principal routing techniques.

The *space domain* utilizes separate waveguides for providing routing diversity, although complexity will be constrained due to physical-layer limitations. This is analogous to the electronic paradigm's utilization of multiple sets of metal wires for providing different network paths. Decisions on how a network message is routed is determined with electronic logic. For photonics, in-flight logic processing of a message is extremely difficult and impractical. Instead, photonics can utilize a separate control network that sets up the optical link a priori (Sect. 5.2), or alternatively uses a store-and-forward technique which requires optical-electronic-optical (O-E-O) conversion to buffer network messages (Sect. 7.1).

The *wavelength domain* uses WDM to accomplish routing. No analogous technique exists within the electronic circuit domain. As described in Sect. 2.1.1, WDM allows the transmission of multiple different wavelengths of light through a single physical waveguide. Routing is enabled by introducing devices that have a behavior which is dependent on the wavelength of light. For instance, in order to properly receive a WDM signal, an array of detectors would need a set of filters to separate the individual wavelength signals so that each can be independently detected (see Sect. 3.6.3).

The *time domain* leverages temporal positioning of messages for routing optical messages. Transmissions of data on the optical network are scheduled such that they do not interfere with each other. Time-arbitrated networks for chip-scale communications are discussed in Sect. 7.1.

5.1.1 Packet-Switching Networks

A review of packet-switching networks is first presented, then contrasted to explain why photonics, differently from electronics, is *not* a good fit to implement this architecture.

The typical functions of a classical electronic packet-switching network node are to perform storage, arbitration, and switching. The term store-and-forward (SaF) is used to describe the nature by which messages travel through a packet-switching network. SaF refers to the process by which network nodes receive and hold data (store) until necessary network resources are ready for data to be transmitted to its destination (forward). The process of determining network-resource availability and allocating them to packets is called arbitration. The act of sending and directing the data is called switching. The SaF maneuver can be utilized multiple times along intermediate nodes in the network in order to deliver a data from source to destination.

Transmission of data first requires that it be broken up into manageable pieces. Each of these pieces of data, referred to as a *payload*, are encapsulated into *packets*. The packet is prefixed with a *header* which can contain information for source location, destination location, and identification. The information contained within the header is used at each network node for determining how and where to send the packet.

The steps for an example SaF process of packet-switching networks is illustrated in Fig. 5.3. Each router contains input buffers, an arbiter, and a switch. The following describes the enumerated steps in Fig. 5.3:

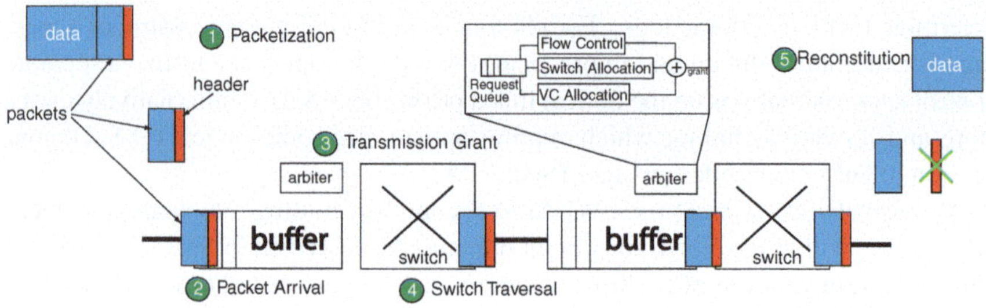

Fig. 5.3 The store-and-forward process of packet-switching networks [2]

Step 1 Transmission of data in a network begins at the source node where packets are generated by segmenting a piece of data into multiple pieces (i.e. payloads) and appending an appropriate header. The header typically contains source address, destination address, and a unique identifier to determine the ordering of packets. The size of the payload is determined by the design of the network.

Step 2 When a packet arrives at a router, it is stored in a buffer and control information is sent to an *arbiter* to determine how it should be routed. The arbiter must first determine the next location within the network that the packet will be sent to. If it is not the destination node, then it will be another router which is described as being *down stream*. The arbiter then must allocate an input and output port on the switch, and may optionally assign a virtual channel (VC).

Step 3 If the arbiter successfully allocates the switch (and optionally a VC) then the switch is configured and the packet is removed from the buffer so it may proceed through the network.

Step 4 After the packet traverses through the switch to the next router, the switch (and VC) resources are released so that a subsequent packet can use them. Steps 1–4 repeat until all data has been transmitted.

Step 5 At the destination, the payload of each packet is ordered according to the unique identifier that was generated by the source. In most instances, the destination node is notified that data has arrived upon completion of the entire message.

Figure 5.4 shows the structure of a typical packet-switching router. For a network that utilizes virtual channels, a virtual channel multiplexer will immediately route an incoming packet to the correct buffer space upon arrival. Transmission requests and grants are sent on buffer control wires, and packets traverse along the data path wires through the switch. This general router structure along with the SaF paradigm has led to innumerable network and NoC architecture improvements over the years, including implementation improvements to lower latency, optimized functions like broadcast, and routing techniques like wormhole routing, cut-through routing, and dynamic routing. Techniques for managing buffer space in the network have been proposed for increased utilization, such as credit and token flow control.

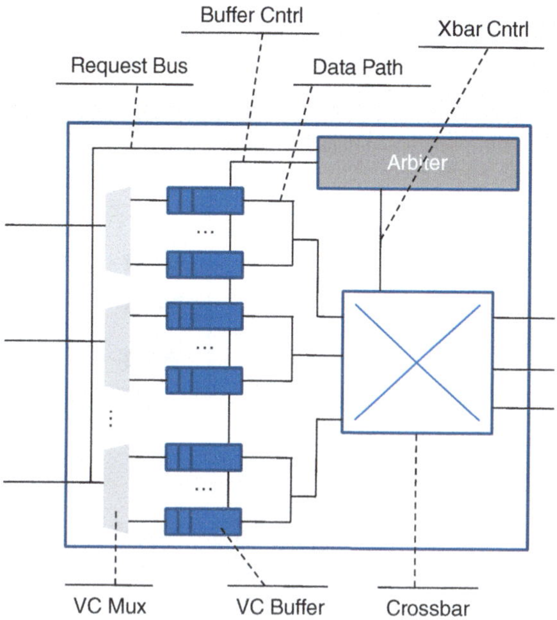

Fig. 5.4 Structure of an electronic store-and-forward packet-switching router [2]

The fundamental difference of photonics is that while it has benefits of high bandwidth and low latency, it lacks the ability to store information for an arbitrary amount of time to allow for arbitration and routing. This limitation prevents a pure photonic system from using the SaF paradigm and the traditional (electronic) router designs.

The remainder of this book will examine several photonic interconnection network designs that circumvent this limitation by utilizing alternative methods for guiding packets towards their destination. Each of these methods adheres to the limitations of optics and ensures a complete path from source to destination.

5.2 Circuit-Switching Basics

Since a key limitation of photonics is the inability to perform inflight buffering and logic, one alternative is to use circuit switching. Circuit-switched networks, also called *spatially*-switched networks, use data lines use data lines with low processing overhead to produce high-bandwidth links, and control packets on a separate control network to establish end-to-end paths before the transmission of data. An important aspect of circuit switching is that once a path is setup, no additional processing is required to ensure a packet reaches its intended destination.

For photonics, this translates to using broadband WDM signaling to produce extremely high-bandwidth connections between communicating nodes by multi-

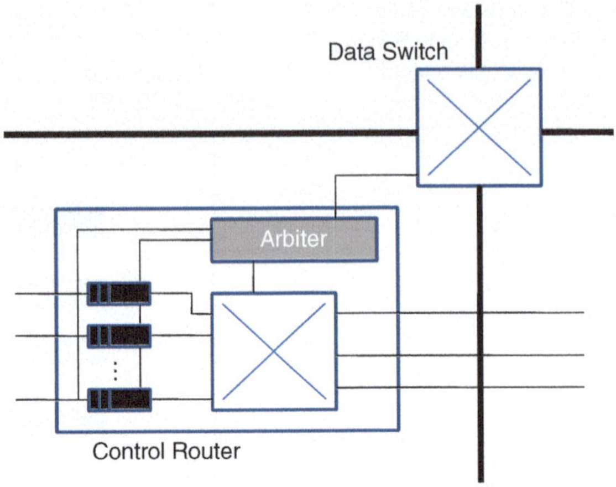

Fig. 5.5 Schematic of a circuit-switching router. The router is composed of a control router and a data switch which are tasked with the flow of control packets and data packets, respectively

plexing data onto many parallel wavelengths. Switching is enabled through the use of broadband switches that can be actively controlled to pass or switch all the wavelengths concurrently, as discussed earlier in Sect. 3.5.2. Since this method requires logical control to activate or de-activate each ring resonator along a path from source to destination, a lightweight electronic packet-switched network can be used to arbitrate access to the network of photonic switches. A path-setup protocol, discussed in Sect. 5.2.1, is used to set up an end-to-end path through the network *before* data is transmitted, followed by transmission of the data when setup is complete.

The overall structure of a *circuit-switching router* is illustrated in Fig. 5.5. The router is composed of two main sub-components: the control router and the data switch. The control router and the data switch handle the movement of control packets and data packets, respectively. Control packets that flow into the router contain information about a data path that needs to be setup through the co-located switch. The control router will send control signals to the data switch accordingly.

The electronic *control router* is composed and functions similarly to the packet-switching router. One key difference is that the control router only handles the transmission of control messages which consists of just the header portion of a normal network packet. Control routers do not have to support large flows of data, therefore are optimized for latency and not throughput. This optimization is referred to as *lightweight*, since the channels can be narrower (few number of wires between routers) and the buffers can be smaller (since only header information needs to be stored). The lightweight *control network* that is created using these control routers is optimized to save power with little or no sacrifice in performance.

The data switch portion of the circuit-switching router contains no routing logic but is typically characterized with low latency and high throughput. The switch is manipulated by the signals that come from the control router. The electronic implementation of this switch would utilize high-speed electronics.

5.2.1 Path-Setup Protocol

The *path-setup protocol* is the mechanism that the control routers use to ensure that the data switches in the network are always set correctly for proper end-to-end data transmission. The path-setup protocol also specifies a procedure for how to resolve conflicts when multiple transmissions require a common network path. This is accomplished with the use of four types of control messages: *path-setup*, *path-ACK*, *path-blocked*, and *path-teardown*. Figure 5.6 shows an example of how the path-setup protocol works. The sequence of events in Fig. 5.6 occur as follows:

Failed Setup. In this scenario, the path-setup protocol is interrupted by a conflict within Router_2.

Step 1.a. The network interface (NIF) is tasked with handling the communication protocol used for transmitting data and is found at every endpoint node in the network. A network transmission is initialized at a NIF which wishes to send data (NIF_{send}) to another NIF in the network ($\text{NIF}_{receive}$).
Step 1.b. When the path-setup message arrives at Router_1, the router will attempt to perform two tasks: route the control packet and provision the circuit path. The creation of the circuit path requires the reservation of data switch resources which typically consists of selecting an input and output port. After the control router has determined the input-output port pair, Router_1 checks that neither input port nor output port is reserved or in use by the data switch (indicated by a *PathFree* flag). Additionally, even if the input and output port of the data switch is free, it is possible that a conflict exists with other ports due to the internal structure of the data switch (this is known as a *blocking switch*). If the internal structure of the switch does not allow for this to happen, then it is termed a *non-blocking switch*. If it is a blocking switch, then the router must perform an additional check to ensure no conflict occurs. In this example, Router_1 is able to forward the path-setup message to Router_2, and Router_1 internally flags that the associated resources of the input-output pair of ports as *PathReserved*, which indicates that the path is in the process of being set up.
Step 1.c. The path-setup message continues to Router_2 where it repeats the procedure that occurred in Step 1.b. In this instance, an internal conflict within the data switch prevents the allocation of the necessary network resources for the circuit path (e.g. the output port has already been reserved for another transmission). Router_2 alters the message by changing it into a *path-blocked* message and setting the destination to the originating source node. No reservation of switch resources is made.
Step 1.d. The arrival of the path-blocked message indicates to Router_1 that it should free the input-output port pair that was reserved in Step 1.b. This release will allow other transmissions to reserve the input or output port. The control message then gets forwarded back to NIF_{send}. NIF_{send} becomes aware that the circuit-path setup failed.

Step 1.e. NIF_{send} can either immediately re-attempt the path-setup protocol or implement a *backoff*. Backoff is used in much the same way as in Transmission Control Protocol (TCP) in the Internet which uses backoff to manage network congestion when establishing a link. Backoff in the circuit path-setup protocol gives the network time to free the contended resources before attempting a new setup request. This has the added benefit of saving power and preventing further congestion.

(Alternative Path Blocked Resolution) Instead of the path-blocked mechanism that began at Step 1.c., the circuit path-setup protocol could stall the path-setup message in the network until the reserved resources are freed. This is problematic for large networks with heavy traffic since it will induce much more contention by holding onto resources. This alternative also requires a mechanism to prevent deadlock in the case that multiple path-setup routines become cyclically dependent on each other's resources.

Successful Setup. In this scenario, the path-setup protocol successfully completes and data is able to be transmitted.

Step 2.a. A new path-setup message is transmitted, and this time it can reserve the required resources at each router along its desired network path.

Step 2.b. The path setup completes upon the arrival of a path-setup message at $NIF_{receive}$. $NIF_{receive}$ sends a path-ACK message back through the network towards NIF_{send}.

Step 2.c. As the path-ACK message is sent through $Router_2$ and $Router_1$, their respective control routers will adjust the flags of the reserved resources to *PathSetup*. At this point, the logic within the control router will actually send the appropriate signals to the data switch to implement the requested circuit path.

Step 2.d. After NIF_{send} receives the path-ACK message, it can now assert that the end-to-end path is available and that data can be safely transmitted towards the destination.

Step 2.e. After the transmission is complete, the path reservation is released by sending a *path-teardown* message through the control network. Figure 5.6 shows the teardown beginning from NIF_{send}, however path teardown can also be initiated from $NIF_{receive}$. It may even be preferable to have the destination initiate the path-teardown message in cases where retransmission is necessary due to bit errors. If the path teardown is initiated from the sending node, it should be sent in such a way that the path-teardown message does not reach each router along the path before the data transmission has finished utilizing it.

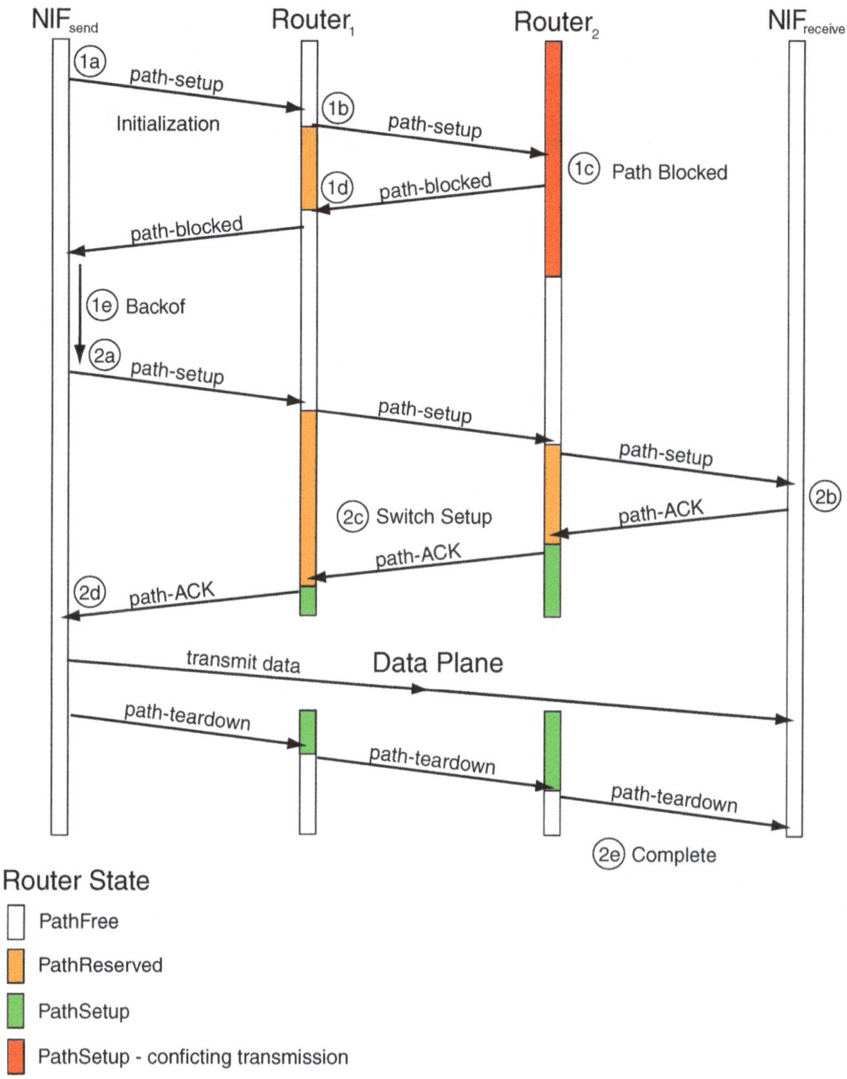

Fig. 5.6 A timing diagram showing an example sequence of control packets for constructing a path through the network

5.2.2 Photonic Spatial Switch Design

Much in the same way that network topology affects latency and congestion in packet-switched electronic networks, the design and placement of photonic switches in a photonic circuit-switched network has consequences on performance. This section describes the design of a photonic circuit-switched network based on a mesh topol-

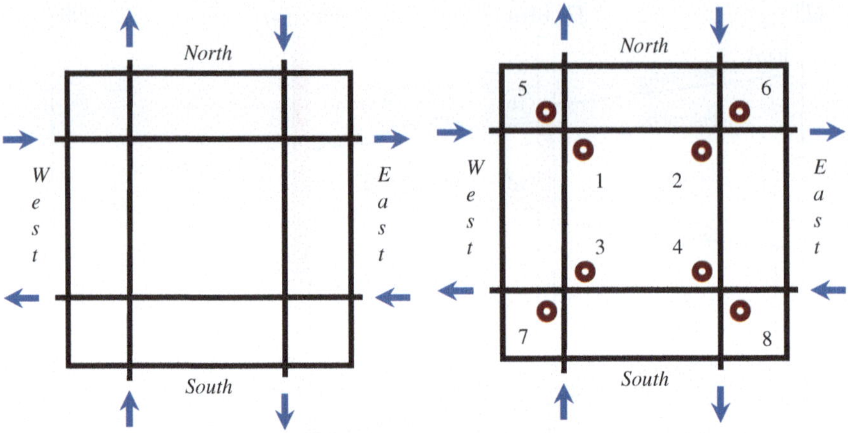

Fig. 5.7 4×4 photonic switch design. **a** With only waveguides. **b** With waveguides and broadband ring resonator switches. Rings are numbered and used to identify in Table 5.1 the path it enables

ogy. This exercise will highlight choices in layout and devices in the context of the photonic topology design space.

The basic building block of a spatially-switched network is the spatial switch. As discussed earlier in Chap. 3, broadband ring switches can be used to simultaneously switch all the wavelengths of a WDM signal. By combining several broadband ring resonator switches together, higher-radix switches can also be realized.

A common network device that can be implemented with ring resonators is a 4×4 broadband switch, with a bi-directional port (input port and output port) for each of the cardinal directions (North, South, East, and West). A naive design is shown in Fig. 5.7a. The construction of this 4×4 switch contains four intersecting waveguides forming a bi-directional crossing. The column labeled Fig. 5.7a in Table 5.1 summarizes the functionality of this simple configuration. This design doesn't implement all possible paths. Paths denoted with a hyphen (–) in Table 5.1 indicate a valid path. Even when presuming that path turnarounds (e.g. *from* the East port *to* the East port) are not required, several input-output path combinations are still missing.

Broadband ring resonators can be added at waveguide crossings to implement dynamically adjustable paths, as shown in Fig. 5.7b, first described by Shacham et al. [3]. The column labeled Fig. 5.7b in Table 5.1 identifies the ring that is used to implement the indicated path. Paths marked with hyphens indicate available paths that do not require the activation of a ring. In this case, all possible paths (excluding turnarounds) are available. Note that eight rings are needed to enable all paths.

Although the switch in Fig. 5.7b can implement all paths, this implementation exhibits blocking characteristics. For example, the path from West to South cannot be active at the same time as the path from North to West due to a conflict in resources. The two paths rely on a common waveguide segment and require conflicting ring states (in rings 2 and 4). This congestion becomes compounded as you consider an entire network built using this switch design.

Table 5.1 Switch design

Input port	Output port	Ring/Path				
		Figure 5.7a	Figure 5.7b	Figure 5.8a	Figure 5.8b	Figure 5.8c
W	N		5	–	–	3
W	E	–	–	4	7	–
W	S		2	8	6	8
N	E		6	2	–	2
N	S	–	–	6	5	–
N	W		4	–	1	5
E	N		3	1	3	1
E	S		8	–	–	6
E	W	–	–	5	2	–
S	N	–	–	3	4	–
S	E		1	–	8	4
S	W		7	7	–	7

This problem motivates the design of non-blocking switches, which guarantees that an optical path can be established whenever the input port and output port are not conflicted. The first design considered here was proposed by Wang et al. [4], and illustrated in Fig. 5.8a. The next two switch designs are shown in Fig. 5.8b, c and were proposed by Chan et al. [5] to reduce total insertion loss in a circuit-switched network. Table 5.1 has columns that specify the rings that implement each path for each of these three non-blocking switch designs.

The switch in Fig. 5.8b, labeled *Symmetric*, is designed to reduce insertion loss by minimizing waveguide crossings. This design takes a naive optimization approach by reducing the average insertion loss for all combinations of input-output paths.

The switch in Fig. 5.8c, labeled as *StraightPath*, was proposed to take advantage of assumptions about dimension-order routed networks by implementing straight default paths (i.e. no switching through a ring). A path is considered a *default path* if the optical path through the switch is enabled without enabling a ring (i.e. dropping through a ring). While it is not always the case, it is typical for a default path to exhibit the lowest insertion loss. In a dimension-ordered network design, a single turn is committed during a network transmission, therefore straight movement through a switch is much more common. This design also presumes that dropping through a ring incurs more insertion loss than passing by a ring or traversing a waveguide crossing.

Higher radix photonic switch designs beyond 4 × 4 are also possible. Figure 5.9a shows a design that is scalable to any number of ports, with the caveat that all input ports are on the same side, as well as the output ports. For simplicity, this design is configured as a 4 × 4 switch. The exclusion of turnarounds paths allows some rings to be removed, as shown in Fig. 5.9b. In Fig. 5.9c default paths are included so that the number of ring switches can be further reduced. This final reduction achieves the minimum number of 8 rings for a 4 × 4 switch. Even though this design contains more

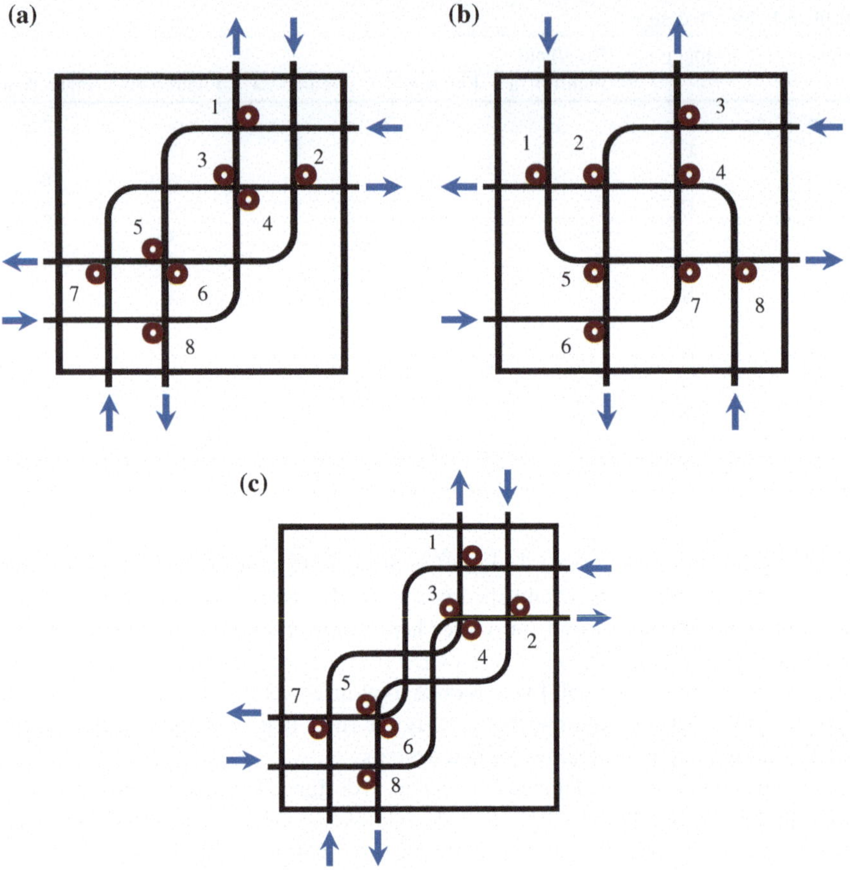

Fig. 5.8 Three 4 × 4 non-blocking switch designs. **a** the original 4 × 4 non-blocking switch, **b** a 4 × 4 non-blocking switch design that minimizes the number of waveguide crossings, and **c** a 4 × 4 non-blocking switch that minimizes insertion loss for the straight path cases (North <–> South, and West <–> East)

waveguide crossings (and thus higher insertion loss) than the switches of Fig. 5.8, it is scalable to an arbitrary number of ports.

5.2.3 Modulator and Detector Banks

An important component of a circuit-switched network is the subsystem that converts network messages between the electrical domain and optical domain. These subsystems are called the modulator bank and the detector bank.

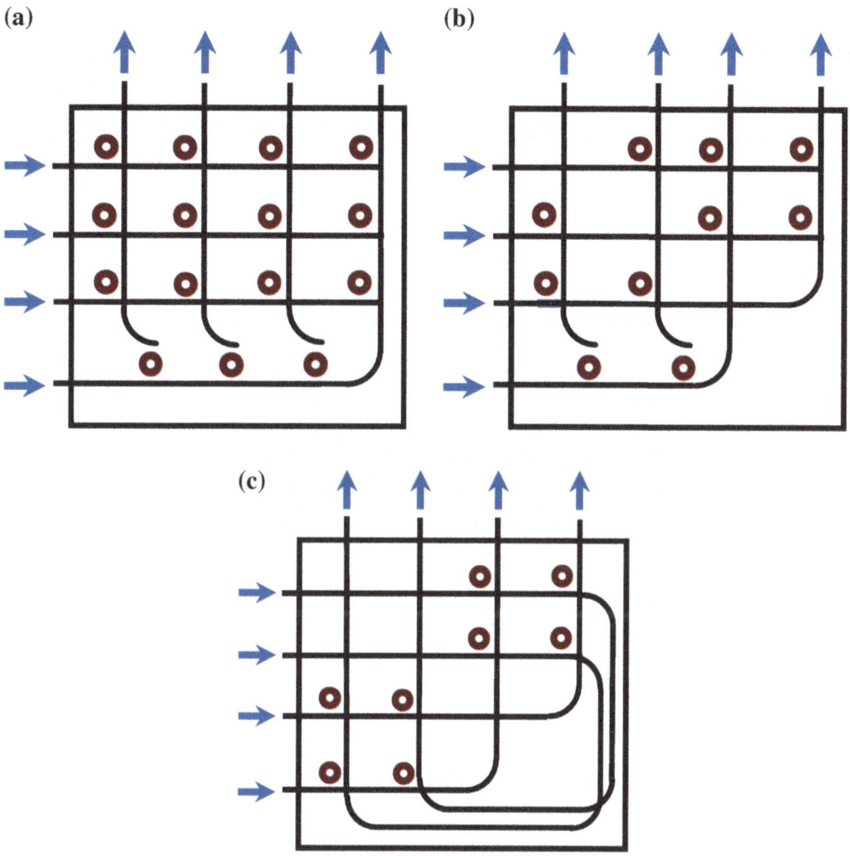

Fig. 5.9 Designs of higher radix photonic switches. **a** Basic grid-style construction of a switch with a scalable number of ports. **b** The assumption that no turnaround paths are required allows the removal of certain rings. **c** Implementation of a default path for each input port enables a further reduction of rings

5.2.3.1 Modulator Banks

Figure 5.10 shows a modulator bank consisting of four ring resonator modulators. The modulator bank translates data signals from the electrical domain into the optical domain. Network messages are first held in NIF buffers (clocked with *clk1*). When the messages are ready to be transmitted, they are serialized with shift registers (clocked with *clk2*) and sent to high-speed drivers, which manipulate the ring modulators. Each ring modulator is designed to have an FSR much larger than the range of wavelengths being using. Also, wavelength channels are spaced a spectral distance of ω_c apart. The spacing must provide enough distance between channels such that the modulators that act on the channels do not perturb other nearby wavelength channels. Each modulator in the modulator bank operates independently, therefore

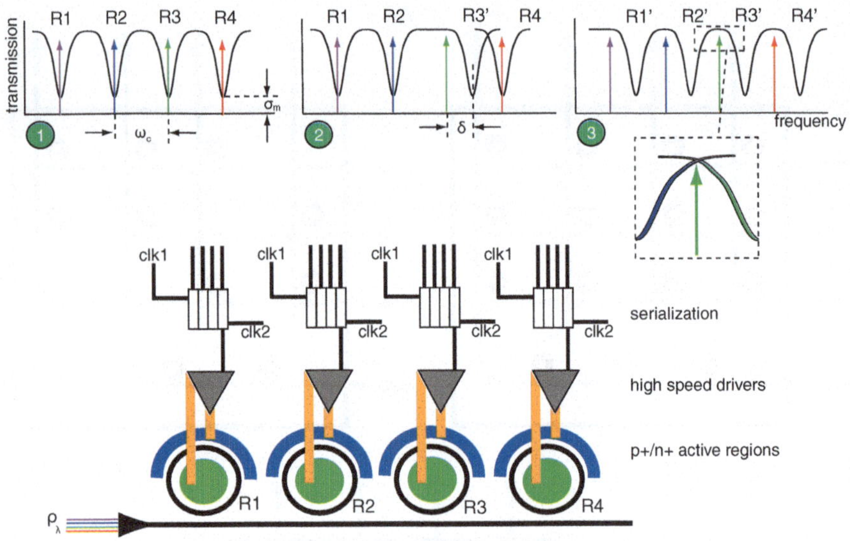

Fig. 5.10 Modulator bank schematic and possible transmission issues. Each *plot* depicts four resonances, each corresponding to one of four ring modulators (R1, R2, R3, and R4). Three scenarios are considered: (*1*) all modulators are off and transmitting '0' bits, (*2*) the R3 ring is on and transmitting a '1' bit while R1, R2, and R4 rings are off and transmitting '0' bits and (*3*) all rings are turned on and transmitting '1' bits

several transmission scenarios exist with unique physical-layer considerations. Three scenarios are considered here.

The first case considers when each modulator is off and transmitting an optical '0' bit. The resonances of the four rings are designed to exactly align with the four wavelength channels that are spaced ω_c distance apart (illustrated in Fig. 5.10(1)). An important consideration here is the leakage past the modulator due to imperfect extinction and wavelength channel misalignment to the resonance. As shown in the figure, points in the spectrum that transmit the signal ('1' bit) have a high transmissivity (e.g. close to 1) while the resonant modes dip to a minimum transmissivity of σ_m. σ_m is the ratio of power that is transmitted during the generation of a '0' bit divided by the input power of the wavelength channel. The power that leaks through is the power of the wavelength channel times σ_m. A figure of merit for the quality of modulation is its *extinction ratio*, which is the power of a '1' bit divided by the power of a '0' bit. A higher extinction ratio is better. This motivates the design of modulators with high extinction ratio and strong alignment of channel wavelengths with resonance wavelength.

The second case, shown in Fig. 5.10(2), depicts one shifted resonance for R3, by an optical distance δ. This is done so that the optical power of the R3 wavelength channel can pass by the ring in order to transmit an optical '1' bit. During a '1' bit transmission, a primary concern is insertion loss, because '1' bits need to be able to overcome the attenuation of the network and to be detectable by the photo receiver.

For this example, the insertion loss of ring R3 is $\zeta_{modulator}(\delta)$, a value dependent on the distance between the resonance and the wavelength of the data channel assigned to R3. The power of the '1' bit at the output of the modulator is the input power of the wavelength channel times $\zeta_{modulator}(\delta)$. Increasing the shifted distance δ results in lower modulation loss, but requires more power (from electro-optic manipulation). Also, the proximity of the adjacent wavelength R4 will restrict the how far the R3 resonance can be shifted. The further right the resonance of R3 is shifted, the lower is the loss on the wavelength assigned to R3, but the greater the impact on R4. There are also loss contributions from the response tails of the other modulators R1, R2, and R4 which are also dependent on the resonance distance to wavelength of the data channel.

Finally, a third case is shown in Fig. 5.10(3) where all four rings (R1, R2, R3, and R4) are shifted such that '1' bits are generated on all channels. Just like in Fig. 5.10(2), the loss experienced by a particular wavelength channel is dependent on its spectral distance from all the other modulator resonances. The modulators are designed to shift a distance of $\delta = \omega_c/2$ such that the resonance shifted falls midway between adjacent channels. The contribution of losses is dominated by the two closest wavelength channels. Therefore the loss experienced by a modulated wavelength channel can be approximated as $2\zeta_{modulator}(\delta)$.

5.2.3.2 Detector Banks

A schematic of a detector bank is shown in Fig. 5.11. The detector bank converts optical data streams into electrical data streams. A passive ring resonator filter and a photo-detector is assigned to each wavelength channel of the system. The ring resonator filter is designed to divert a specific channel from a WDM signal. High speed receivers and deserializers then convert the high-speed serial optical bits (clocked at *clk2*) to the lower-speed electrical domain (clocked at *clk1*).

One important issue is determining *clk2* which is then used to sample the incoming converted data stream. The frequency and phase of *clk2* need to be adjusted such that every bit of data can be correctly received. This task is known as clock and data recover (CDR). Since networks utilized in CMPs are sensitive to latency, it is paramount that CDR latency overhead be minimized. As of the writing of this book, no practical design exists for recovering a clock with reasonable delay for high speed data in chip-scale photonic interconnection networks. Burst-mode receivers today use phase-locked loops (PLLs) which require microseconds of locking time in theory, and milliseconds in practice.

Figure 5.11(R1) depicts the WDM signal and drop-port response at the first ring R1. Two important considerations are illustrated in the inset: transmission and filter leakage in Fig. 5.11(A) and crosstalk in Fig. 5.11(B). The power of P_λ represents the power of wavelength channel λ (where λ equals 1, 2, 3, or 4) at the input to the detector bank. The filter leakage (σ_f) represents the fraction of the channels power that is not transmitted to the drop port. $1 - \sigma_f$ represented the power of the R1 wavelength channel that is transmitted to the detector, assuming an ideal

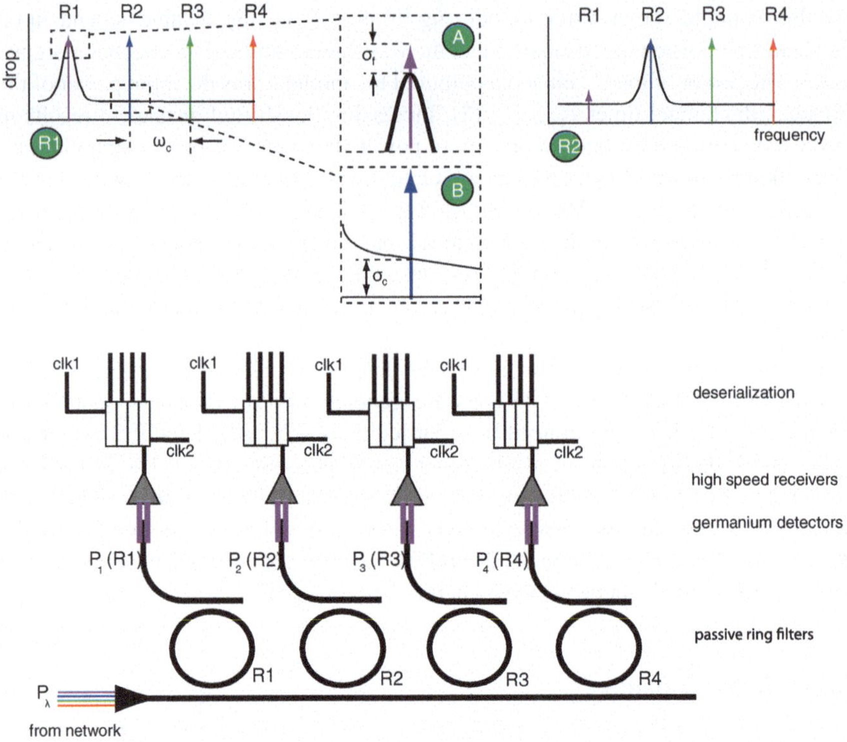

Fig. 5.11 Detector bank schematic and response issues. (*R1*) Wavelength channels at the input of ring R1 and the spectral response at the R1 drop port, and (*R2*) wavelength channels at the input of ring R2 and the spectral response of the R2 drop port. The inset figures: **A** depicts the ratio of power transferred to the R1 photo-detector, and **B** depicts the crosstalk power extracted from an adjacent wavelength channel not intended for R1

lossless ring filter. This leakage is caused by a non-ideal filter response and resonance misalignment. This leaked power is transmitted to subsequent filters in the detector bank and acts as a noise source. Wavelength crosstalk, σ_c, are power contributions from adjacent wavelength channels that leak into detectors (e.g. wavelength R2, R3, and R4 leaking into the detector intended for R1). This leakage is treated as noise since it is not intended and is a function of its spectral distance from the resonant mode. The inset in Fig. 5.11(B) shows the channel intended for R2 leaking into the R1 detector. Crosstalk contributions also come from the wavelength channels assigned to R3 and R4. The R1 detector approximately receives:

$$P_1(R1) = P_1(1-\sigma_{f1}) + P_2\sigma_c(\omega_c) + P_3\sigma_c(2\omega_c) + P_4\sigma_c(3\omega_c) \qquad (5.1)$$

where σ_c is the fraction of power that drops a ring in the form of crosstalk and ω_c is the wavelength channel spacing. This approximation assumes that each ring has the exact same shape but spectrally shifted response. This indicates that the detector will receive the power of the incoming R1 wavelength minus the leakage power, plus crosstalk from the R2, R3, and R4 wavelengths.

Considered next is what happens when the signal is received by R2 (Fig. 5.11(R2)). Crosstalk contributions to R2 begin with the optical power leaked through from R1. Similarly, since some power from the R2 wavelength was leaked into R1, its power is slightly diminished. The detector at R2 can expect to receive:

$$\begin{aligned} P_2(R2) = {} & P_2(1-\sigma_c(\omega_c))(1-\sigma_{f2}) + P_1(\sigma_{f1}\sigma_c(\omega_c)) \\ & + P_3(1-\sigma_c(2\omega_c))(\sigma_c(\omega_c)) + P_4(1-\sigma_c(3\omega_c))(\sigma_c(2\omega_c)) \end{aligned} \tag{5.2}$$

This indicates that the R2 detector will receive the power of wavelength intended for R2 (P_2) diminished by the R1 crosstalk and R2 leakage, plus crosstalk contributions from the other three channels. This motivates the design of filters with low leakage, narrow passbands, and wide channel spacing to decrease crosstalk.

5.2.4 Mesh Topology

Large network layouts benefit from tiled patterns such as a mesh due to advantages in implementation, fabrication, packaging, test, verification, and fault tolerance. The regularity of a tiled design enables easy repetition and scalability.

Figure 5.12 shows the layout of a photonic mesh circuit-switched topology which possess a photonic data network plane and an electronic control network plane. The photonic plane contains a grid of 4 × 4 switches. Each switch has an associated modulator bank and detector bank on the East and South port for injection and ejection to and from the network. The paired modulator bank and detector bank are referred to as an *access point*, which is the location where processing nodes transmit and receive messages.

The interface logic in the NIF is shown in Fig. 5.13 and contains control logic necessary to send and receive optical data by mediating between the local access point and the electronic control network. This basic NIF can only serve one send request and receive one transmission at a time by virtue of the circuit-switching architecture. Changes to this NIF design are discussed later, including selective transmission in Sect. 5.4.2 and time multiplexing in Sect. 7.1.

5.3 Physical Layer Analysis of Photonic Circuit Switching

This section utilizes the physical-layer analysis tools described in Sect. 4.1.4 to examine the scalability and performance of mesh-style photonic circuit-switching topologies. The importance of physical-layer analysis of photonic architectures is exem-

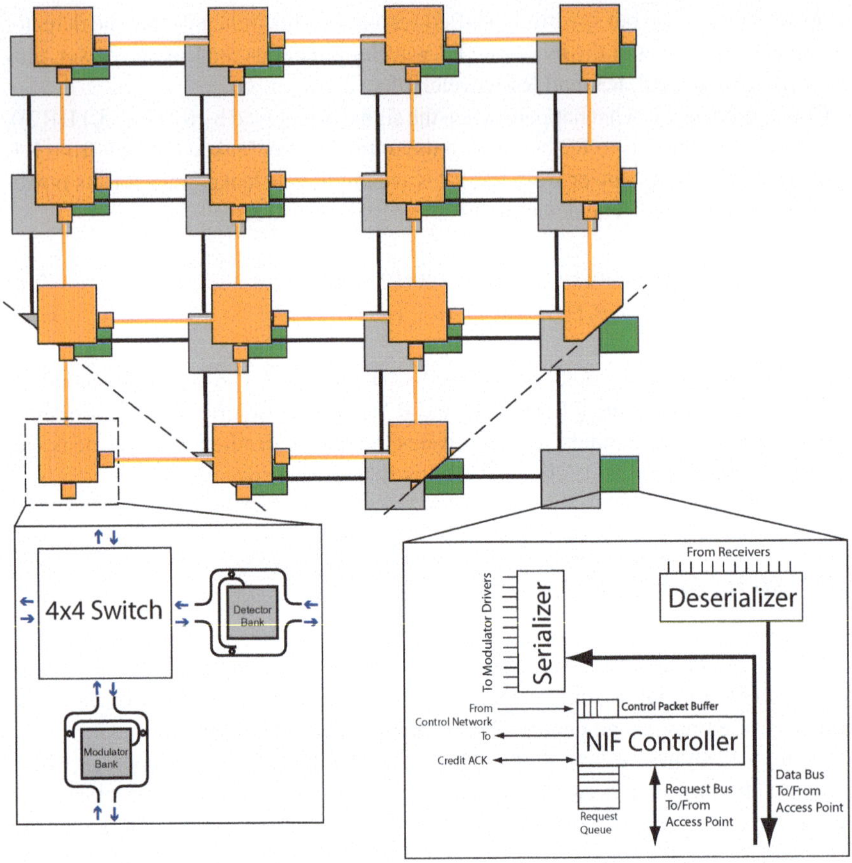

Fig. 5.12 4 × 4 Mesh topology, showing a 5-port photonic switch and NIF block diagram. Key—photonics (*orange*), electronics (*gray/black*), NIF (*green*)

plified by the immaturity of the technology. The chip-scale photonic architecture community proposes many architectures using the novel photonic devices without the critical understanding of physical feasibility. The analysis of the physical properties such as insertion loss and crosstalk is necessary to obtain a realistic design.

5.3.1 Insertion Loss Analysis of 4 × 4 Switch Designs

In mesh-style photonic circuit-switching networks, the 4 × 4 switch provides the main routing mechanism for guiding lightwaves at each intersection of the network. An analysis is conducted to compare the performance of the varying 4 × 4 non-blocking switch designs shown in Fig. 5.8.

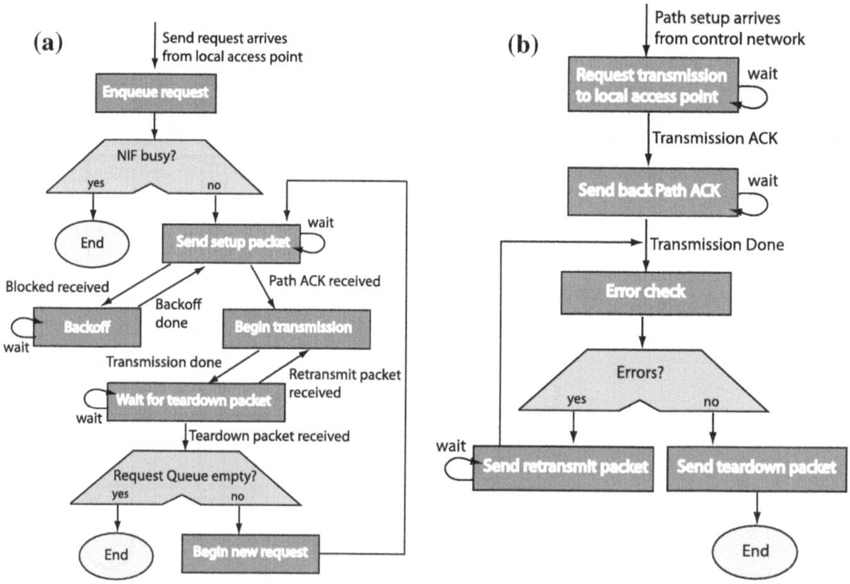

Fig. 5.13 Control flowchart for a circuit-switched NIF. **a** Data-send control; **b** Data-receive control

For this analysis, a circuit-switching folded torus topology is used [6]. The modulator banks, detector banks, and switches throughout the network (Fig. 5.14) use component designs previously described in Sect. 5.2. Additional broadband switches are placed in the network to allow packets to enter (injection) and exit (ejection) the interconnection network topology. Figure 5.14 shows the main folded torus in thick black lines. An additional set of photonic waveguides are included called the gateway access network (shown as thin red lines) and is required to enable entering and exiting the main folded torus network.

The simulation model assumes a tile size of 2.0 mm × 1.5 mm, which is the size of a single core in the Intel 80-core Teraflop chip [7]. Figure 5.15 shows a typical layout of a tile in the torus photonic plane. Each tile consists of a gateway switch, injection switch, ejection switch, 4 × 4 non-blocking switch, and several optical paths to form the torus and gateway access network. Observe that the 4 × 4 switch is customizable with any of the designs shown in Fig. 5.8.

The network uses X-Y dimension-ordered routing. The simulation uses uniformly distributed transmission requests with exponentially distributed interpacket spacing. Although insertion loss is independent of network congestion, an arbitrary constant message length of 50 ns is used for simulation purposes. Layout differences and losses due to the on-chip routing of continuous-wave light and off-chip messages into each gateway are ignored for this simulation.

The insertion loss parameters are shown in Table 5.2. The values are obtained from reported devices and predictions for future scaling.

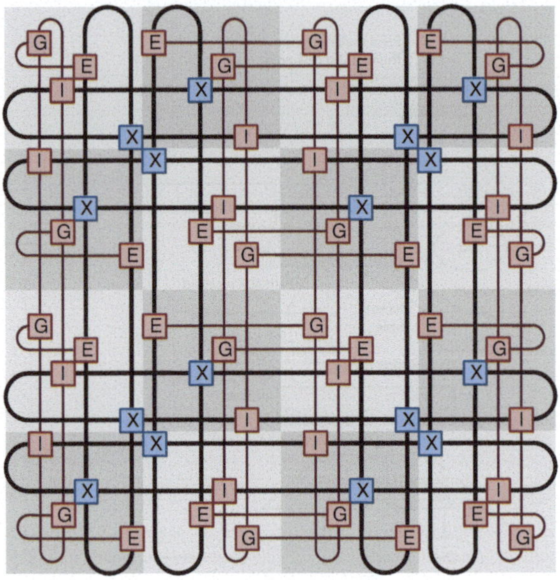

Fig. 5.14 Structure of a 4×4 folded torus topology. The waveguides that make up the torus network are shown as *thick lines*, and the gateway access network for injecting packets to and ejecting packets from the network shown as *thin lines*. The *blocks* represent the following: gateway switch (*G*), injection switch (*I*), ejection switch (*E*), and a 4×4 non-blocking switch (*X*) [5]

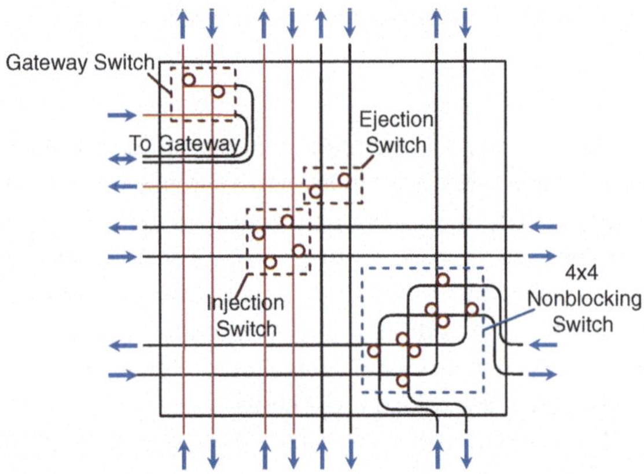

Fig. 5.15 Layout of a tile in the torus network. This includes the type 'Original' version of the 4×4 non-blocking switch shown in Fig. 5.8 [5]

5.3.1.1 Simulation Results

The analysis here investigates how insertion loss is affected by changes in topology size and different switch designs. Tori of size 4×4, 6×6, and 8×8 are considered.

Figure 5.16 shows the distribution of insertion loss that a packet will experience when propagating from source to destination. Minimum losses for each switch layout remains constant for differing network sizes. For every additional two nodes in each dimension, 3.89 dB, 3.66 dB, and 3.36 dB of loss is added to the maximum loss for switch design A, B, and C, respectively. This is a result of the fact that the minimum length path from any two nodes remains the same while the maximum length changes

Table 5.2 Insertion loss parameters—4 × 4 non-blocking switch study

Parameter	Value	Ref.
Propagation loss (silicon)	1.5 dB/cm	[8]
Waveguide crossing	0.05 dB	[9]
Waveguide bend	0.005 dB/90°	[8]
Drop into a ring	0.5 dB	[10]
Pass by a ring	0.005 dB	[10]

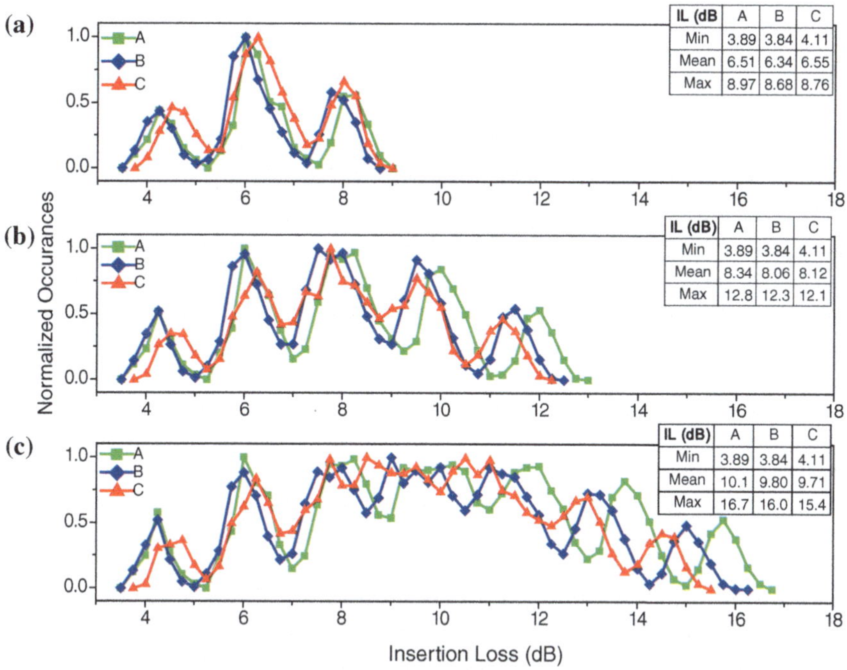

Fig. 5.16 Insertion loss distribution for folded torus topologies of size (**a**) 4 × 4, (**b**) 6 × 6, and (**c**) 8 × 8. Each graph contains plots of three differing switch designs. The *colored plots* correspond as follows: *green* for the **Original** design (Fig. 5.16a), *blue* for the **Symmetric** design (Fig. 5.16b), and *red* for the **StraightPath** design (Fig. 5.16c). *Inset* within each graph is a table of minimum, mean, and maximum insertion losses observed for each switch/topology combination [5]

with number of nodes. The three, five, and seven peaks that appear in the distribution for the 4 × 4, 6 × 6, and 8 × 8 node torus networks, respectively, equate to the maximum number of 4 × 4 non-blocking switches an optical packet must travel through, which rises as the node count scales up.

Each plot in Fig. 5.16 shows the general trend of the different switch designs. The Symmetric 4 × 4 Switch (B) has both a lower maximum and lower minimum loss, in comparison to the Original 4 × 4 Switch (A), as expected. The StraightPath 4 × 4 Switch (C) consistently has a higher loss for the lower bound of the distribution.

Although (C) exhibits higher maximum loss in the 4 × 4 node network than (B), it shows lower loss at sizes of 6 × 6 nodes and higher. This is attributed to the fact that even though the minimum insertion loss for this 4 × 4 switch design is higher than the others, the straight path (from north to south, east to west, or vice versa) has a lower loss because no rings are encountered. In contrast the paths in (A) and (B) that do not pass through any ring resonators implement a turn in dimension. The performance improvement achieved by switch (C) is a consequence of the characteristics of dimension-ordered routing. With dimension-ordered routing, and optical signal will predominately propagate straight through switches and only make a maximum of one turn for any possible end-to-end path. The switch (C) design takes advantage of this property.

5.3.2 Scalability Study Using Physical-Layer Analysis

This section focuses on the physical-layer analysis of space-switched photonic networks. Two previously proposed topologies are the Torus [6] and a Non-blocking Torus [4], shown in Fig. 5.14 and Fig. 5.17, respectively. A node (marked X) is defined as the logical switching point on the network, whereas an access point (marked G) is a gateway where a network user (e.g. a processor node) can initiate or receive a transmission. The nodes are implemented with the non-blocking 4 × 4 switch. The primary folded-torus path in both networks is illustrated with thick lines to represent two waveguides forming a bi-directional link. The remaining thinner lines and blocks (I, E, and S) indicate the location of additional waveguides and switches that compose the gateway access network, which is needed to enter and exit the tori.

Both the Torus and Non-blocking Torus are similar in structure since both networks possess a folded-torus connectivity utilizing 4 × 4 non-blocking switches. However, a key difference is the manner in which access points are mapped to nodes. The Torus has an access point mapped to every node, while the Non-blocking Torus is limited to two access points on each row and column of nodes in the torus in order to achieve a strictly non-blocking network. For example, an 8 × 8 torus would allow 64 access points in a normal configuration, but would only allow 16 access points in a non-blocking configuration. This non-blocking characteristic is possible since the node restriction prevents more than two communications from concurrently traveling on any particular row or column within the network. Since each row or column is actually a bi-directional loop, two separate data streams could travel on the loop (one clockwise, another counter-clockwise) without interfering with each other. Previous studies have shown that the non-blocking property can be advantageous in both throughput and latency compared to blocking networks [4], but performance improvements will be offset by the physical layer constraints that have not previously been considered.

The networks are simulated using PhoenixSim. The simulation topology models assume die sizes of 2.0 × 2.0 cm.

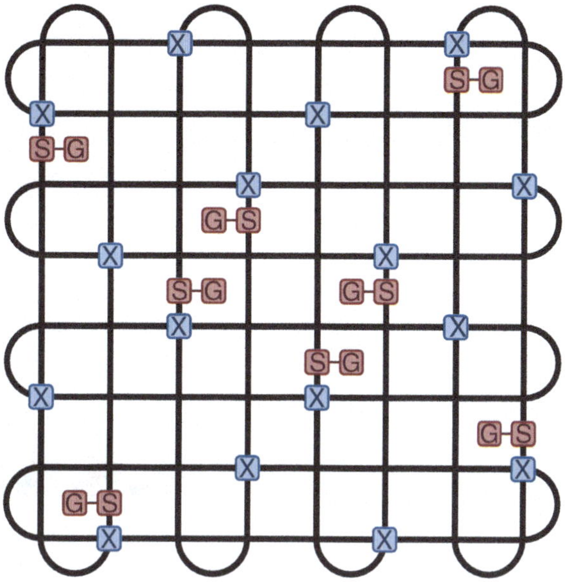

Fig. 5.17 4 × 4 Non-blocking Torus with 8 access points. *X* labels mark 4 × 4 non-blocking switching points. *G* labels mark access points. *S* labels indicate combined injection-ejection switching points [11]

Table 5.3 Insertion loss parameters—photonic circuit-switching analysis

Parameter	Value	Ref.
Propagation loss (silicon)	1.5 dB/cm	[8]
Waveguide crossing	0.15 dB	[9]
Waveguide bend	0.005 dB/90°	[8]
Drop into a ring	0.5 dB	[10]
Pass by a ring	0.005 dB	[10]

5.3.2.1 Insertion Loss Analysis

This study assumes loss parameters close to currently realizable values and are summarized in Table 5.3. Note that ring resonators exhibit a strong thermal dependency which could potentially cause additional losses, increased crosstalk, and disruptions in the network. Thermal management of ring resonator devices is currently an active research topic with proposed solutions that include integrated heaters for thermal compensation [12] and athermal devices [13]. For this simulation discussion, an adequate mechanism for managing this thermal issue is assumed to exist.

The maximum possible loss (across all paths) that a message will incur from each type of component in the Torus and Non-blocking Torus is shown in Fig. 5.18 for networks ranging from 4 × 4 to 18 × 18 nodes. Losses due to bending waveguides and passing by an off-resonance ring are negligible and are not shown. As the photonic network topology scales to support more access points, signals will incur higher losses due to more waveguide crossings and switching elements.

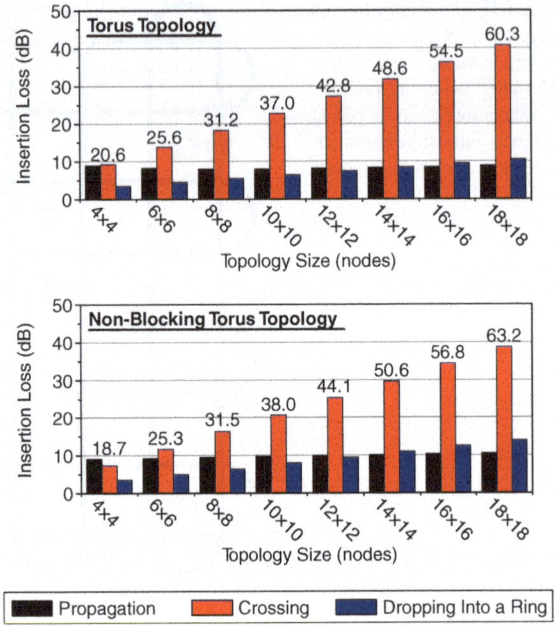

Fig. 5.18 Maximum possible network-level insertion loss by component for varying sizes the Torus and Non-blocking Torus using the parameters listed in Table 5.3. Labeled values represent the peak cumulative insertion loss (in dB) for the network [11]

The waveguide crossings are shown to be the most significant component of optical losses reaching as high as 68 % for the Torus and 61 % for the Non-blocking Torus. The contribution of loss from dropping into a ring on resonance for the Torus and Non-blocking Torus regardless of topology size are approximately 17 % and 20 %, respectively, whereas propagation losses in the 4×4 configuration are as high as 43 % and 49 %, respectively, and gradually decrease in percentage as the topology size increases. The decreasing trend in percentage for propagation loss is due to the assumed fixed size of the die keeping the approximate maximum propagation distance equal while other components continue to scale in number as the topology size increases. Passing by off-resonance rings and passing through waveguide bends induce relatively negligible losses in these topologies. Consequently, the most beneficial improvements to these networks can be achieved through either a reduction of waveguide crossing losses or through the redesign of the switching fabric layout to reduce the number of crossings.

5.3.2.2 Photonic Switching Element Designs

The previous analysis of network-level insertion loss of the Torus and Non-blocking Torus suggests that research advancements in lower-loss crossings will have the most impact in increasing system performance. In particular, two system parameters stand to gain with improvements in loss: the bandwidth available to each access point which is specified by the number of wavelengths, and the number of access points

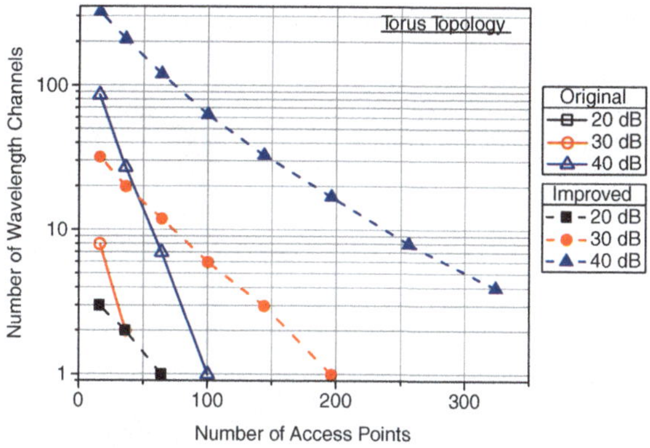

Fig. 5.19 Upper limits on the number of wavelength channels allowed for a given number of access points assuming various network-level optical power budgets in the Torus topology. *Solid lines* assume all realistic parameters (original) and *dashed lines* assume a hypothetical improvement in crossing loss (improved) [11]

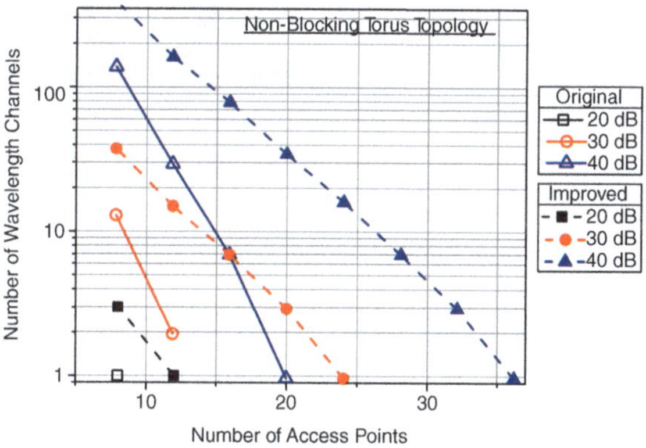

Fig. 5.20 Upper limits on the number of wavelength channels allowed for a given number of access points assuming various network-level optical power budgets in the Non-Blocking Torus topology. *Solid lines* assume all realistic parameters (original) and *dashed lines* assume a hypothetical improvement in crossing loss (improved) [11]

available in the network. Simulated next is a hypothetical improvement in crossing loss, and the use of Eq. 4.9 to determine its impact on network scalability.

Figures 5.19 and 5.20 shows the maximum number of wavelengths that are allowed for varying topology sizes and the change in performance when assuming a hypothetically better crossing loss of 0.05 dB (compared with 0.15 dB in the original case). The gains in system-level performance from the improved crossings are apparent from the network's support for more access points and greater numbers of wavelengths. For instance, assuming a 30-dB allowed network-level optical power budget, the maximum connectivity supported on the Torus scales from 36 access points when using the original crossings to 196 access points when using the improved crossings (a more than five-fold increase). Similarly, the Non-blocking Torus scales from 12 to 24 access points. On the other hand, the Torus topology can be fixed to 36 access points and have a gain in the number of possible wavelength channels from 2 to 20 (ten-fold increase in bandwidth), while a Non-blocking Torus with 12 access points will increase from 2 to 15 wavelength channels. For the case of the Torus network operating with a 20-dB optical power budget and original parameter set, the network

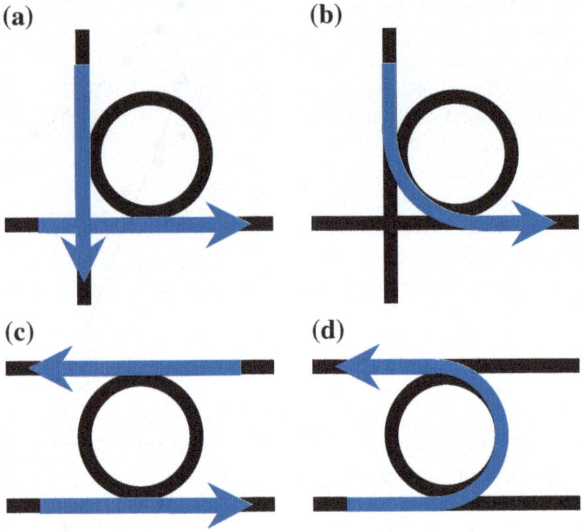

Fig. 5.21 Light propagation in 1 × 2 PSE. **a** Off-resonance propagation with crossing. **b** On-resonance propagation with crossing. **c** Off-resonance propagation without crossing. **d** On-resonance propagation without crossing [11]

configuration is unable to produce any wavelengths since the worst-case insertion loss exceeds the optical budget.

5.3.2.3 Topology Exploration

Network performance improvement can also be achieved though design optimizations that decrease network-level insertion loss. As discussed in Sect. 5.3.2.1, waveguide crossing losses are the dominant contribution to the total optical insertion loss. Therefore, designs that decrease the number of crossings will be advantageous. TorusNX and Square Root were designed with this objective in mind.

A significant amount of loss in the original Torus is attributed to two reasons. First, the usage of the gateway access network introduces an additional set of waveguide crossings which produce a high insertion loss overhead. Secondly, the Torus (and also Non-blocking Torus) is designed using only the 1 × 2 and 2 × 2 PSEs which both contain an embedded waveguide crossing (Fig. 5.21a, b shows the 1 × 2 case, Fig. 5.22a, b shows the 2 × 2 case). These switch designs were suitable for prior investigations into photonic networks since the studies did not consider insertion loss [6], but more recent studies have shown that the overall system performance would be significantly impacted [11]. In many circumstances, a designer can take advantage of alternative 1 × 2 (Fig. 5.21c, d) and 2 × 2 (Fig. 5.22c, d) PSE designs, which eliminate the crossing and reduce the insertion loss impact on off-resonance message traversal but keep similar switching functionality.

The TorusNX topology (Fig. 5.23) is designed to preserve the connectivity and scalability of the original Torus topology while lowering the overall insertion loss. The name of this topology means 'torus, no crossings' and alludes to the strategy

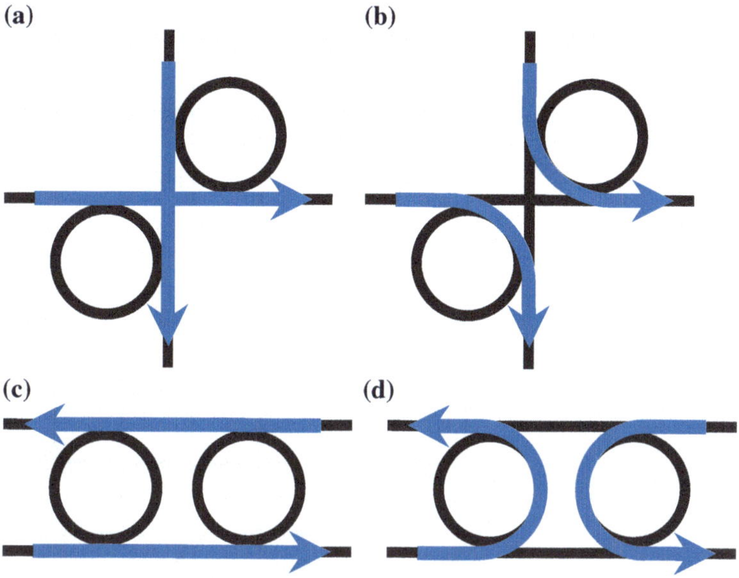

Fig. 5.22 Light propagation in 2 × 2 PSE. **a** Off-resonance propagation with crossing. **b** On-resonance propagation with crossing. **c** Off-resonance propagation without crossing. **d** On-resonance propagation without crossing [11]

used in the designing of this network. Many improvements in device arrangement were made in order to significantly reduce waveguide crossings and to reduce the insertion loss overhead. In contrast with the Torus which required a complex gateway access network to facilitate injection and ejection from the network, TorusNX uses a new gateway design (Fig. 5.24) which splits the access point into separate blocks for modulation and detection and circumvents adding any additional crossings to the torus through the use of the 1 × 2 PSE variant. The modulation block enables a message to be injected into north or south directions while the detection block can receive signals coming from the east or west direction. This scheme is well suited for dimension-order routing which is the assumed routing policy for this topology. TorusNX also uses the optimized StraightPath version of the 4 × 4 non-blocking switch shown in Fig. 5.8c to perform better in dimension-order routed topologies.

The Square Root topology (Fig. 5.25) was also designed with fewer waveguide crossings and fewer switches in mind by simplifying the entire network into only using 4 × 4 non-blocking switches. In addition to the axioms used to reduce insertion loss in the physical layer, the Square Root also uses hierarchical organization to simplify routing, and path multiplicity between organizational units to increase performance.

The Square Root topology is constructed recursively beginning with a 2 × 2 *quad*, shown in Fig. 5.25a, which has no waveguide crossings outside the 4 × 4 switches. A 4 × 4 Square Root is composed of four sets of quads, and is shown in Fig. 5.25b,

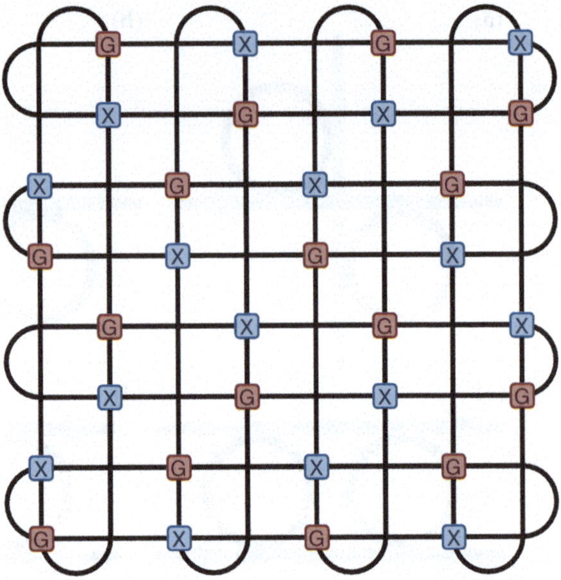

Fig. 5.23 4 × 4 TorusNX network with 16 access points [11]

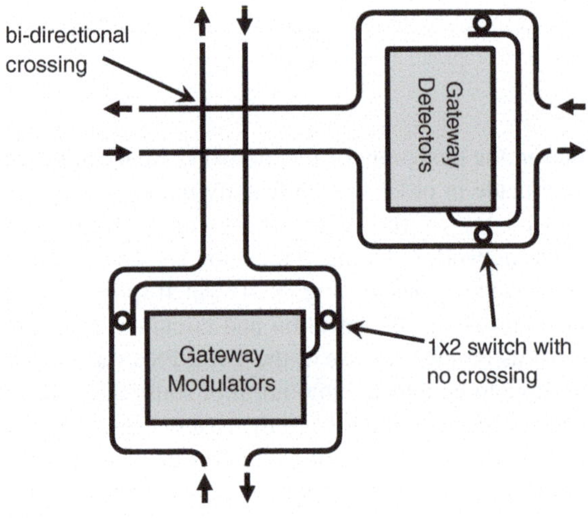

Fig. 5.24 Design for a photonic gateway with an integrated bidirectional crossing [11]

connecting quads through central switches and inter-quad express lanes. In a similar fashion, an 8 × 8 Square Root can be constructed from four 4 × 4 Square Roots. This recursive construction can be used to build any size square topology with dimensions equal to any positive integer power of two.

The insertion loss performances of TorusNX and Square Root assuming realistic loss parameters are shown in Fig. 5.26. For the radixes examined, TorusNX has between 23 and 29 % lower network-level insertion loss in comparison to the original Torus, while Square Root has between 31 and 46 % lower loss. In the case of 8 × 8

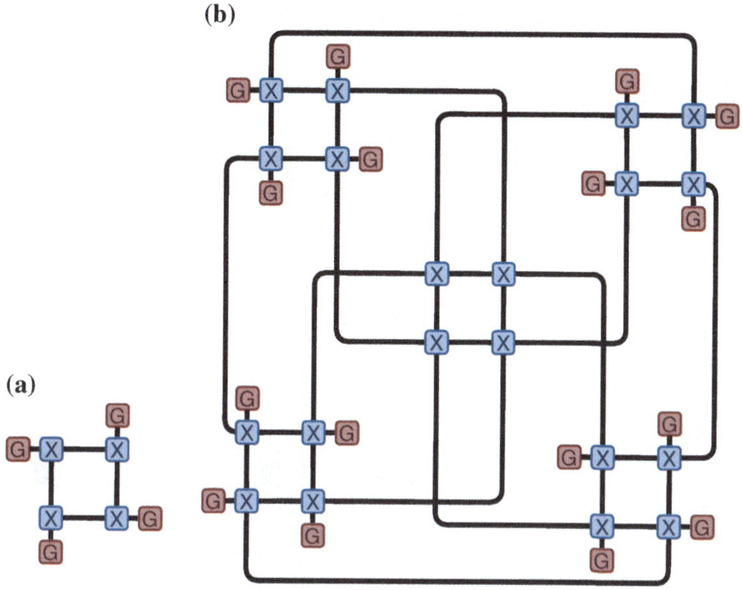

Fig. 5.25 **a** The basic unit of the Square Root topology, a 2 × 2 quad. **b** A 4 × 4 Square Root [11]

topologies, the Torus contains 3200 waveguide crossings, while TorusNX reduces this number to 1796, and Square Root further reduces it to 1080. As before, improved crossing loss can also be applied to these designs to further improve the scalability and performance (Figs. 5.27 and 5.28). Assuming the same 30-dB optical budget and improved crossing losses, both networks are able to achieve the maximum size network simulated in this study (324 access points for TorusNX, 256 access points for Square Root). The remaining optical budget of both maximally-scaled networks allows the transmission of seven wavelength channels.

The results of this insertion-loss analysis clearly indicate that the newly developed networks are better in sustaining higher bandwidths and more access points for better overall system performance. However, for a fixed network design, optical power budget, and device performance, determining the optimal number of wavelengths and access points to use will largely depend on the specific system requirements being targeted. As an example, we can choose to maximize the total ideal network throughput (number of access points × number of wavelengths per access point × data rate per wavelength) of the TorusNX topology. A 30-dB optical power budget, improved device parameters, and a 10-Gbps modulation rate per wavelength are assumed. At one extreme, selecting the maximum number of access points (324) while using a single wavelength achieves a throughput of 22.6 Tbps. On the other hand, maximizing the number of wavelengths (70) would allow a total of 16 access points which results in a throughput of 11.2 Tbps. A balance of the two parameters, in

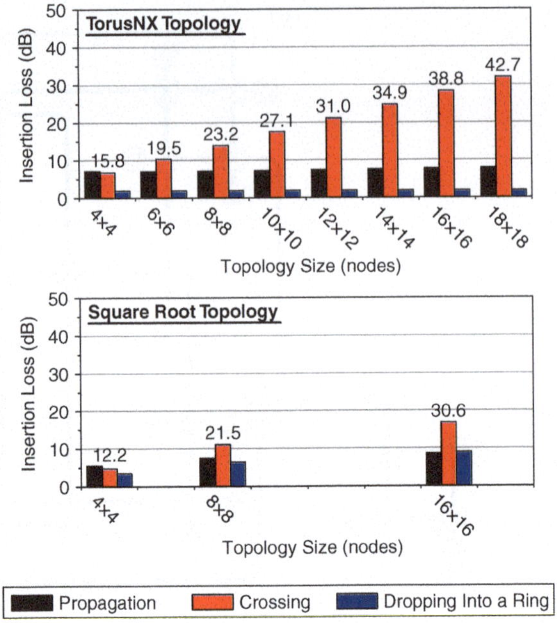

Fig. 5.26 Maximum possible network-level insertion loss by component for varying sizes of TorusNX and Square Root using the parameters listed in Table 5.3. Labeled values represent the peak cumulative insertion loss in dB [11]

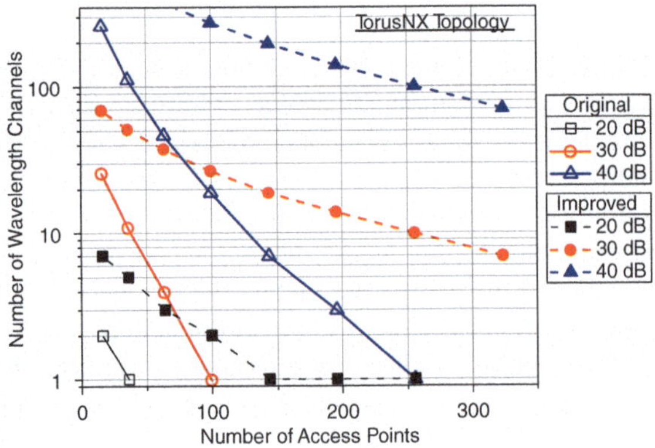

Fig. 5.27 Upper limits on the number of wavelength channels allowed for a given number of access points assuming various network-level optical power budgets in the TorusNX topology. *Solid lines* assume all realistic parameters (original) and *dashed lines* assume a hypothetical improvement in crossing loss (improved) [11]

fact, achieves the best throughput performance at 27.4 Tbps when using 196 access points with 14 wavelengths.

5.3.2.4 Crosstalk Analysis

For system performance, it is useful to report the SNR, which is a measure of the integrity of the message being transmitted. The signal power is calculated based on the injected power and the network-level insertion loss, while the noise power

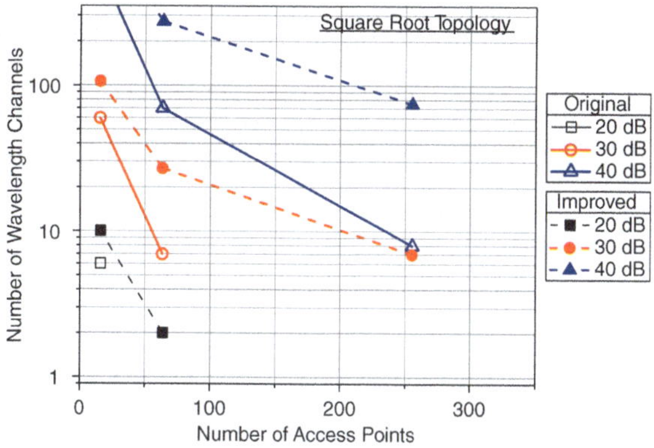

Fig. 5.28 Upper limits on the number of wavelength channels allowed for a given number of access points assuming various network-level optical power budgets in the Square Root topology. *Solid lines* assume all realistic parameters (original) and *dashed lines* assume a hypothetical improvement in crossing loss (improved) [11]

is derived from the several sources outlined in Sect. 4.1.4.2. The crosstalk analysis assumes non-WDM (single-wavelength) transmission, therefore N_{intra} is set equal to zero. This analysis only considers laser noise and inter-message crosstalk. For this reason the presented results can be thought of as an upper bound in OSNR performance.

Determination of laser noise is based on laser and modulator performance. For continuous-wave quantum cascade laser, RIN has been measured to be about -150 dB/Hz for a 10-mW output [14]. Silicon ring modulators have been demonstrated with extinction ratios of about 9 dB when modulated at 12.5 Gbps [15]. Polysilicon ring modulators have also been demonstrated with extinction ratios of 16 dB during DC operation, and 10 dB with active signaling at 2.5 Gbps [16]. The laser noise power can be calculated from Eq. 4.10 because the signal power is known.

The crossings and ring switches are the main contributors of inter-message crosstalk. Waveguide crossing leakage has been measured at -40 dB below the signal power [9]. Similarly, the ability of a ring to resonate or pass a particular optical wavelength channel is also non-ideal. A signal that is on-resonance with the ring will mostly drop through the ring, while a small portion of the optical power will continue through in the off-resonance direction. The same is true in the case of an off resonance signal, which will partially leak onto the on resonance direction. This small leakage of the optical signal can interfere with other propagating messages by contributing crosstalk noise. This behavior is characterized by the extinction ratio, which has been measured experimentally to be 28.6 dB for the through port and 18.7 dB for the drop port [10]. All noise related parameters for the crosstalk analysis are listed in Table 5.4.

The OSNR measurements for the four networks are reported in Fig. 5.29 for varying message sizes. Communications on space-routed topologies have varying ratios of photonic activity to electronic activity due to the separate electronic control and photonic data planes. Network activity exclusively takes place on the control plane during the provisioning and release stages of a photonic path. Therefore no optical signal is injected during these periods. As the transmission message sizes increases,

Table 5.4 Crosstalk and noise parameters—photonic circuit-switching analysis

Parameter	Value	Ref.
Lasers (RIN)	−150 dB/Hz	[14]
Modulation (modulation index)	16 dB	[16]
PSEs—through port extinction ratio	25 dB	[10]
PSEs—drop port extinction ratio	20 dB	[10]
Crossings (crosstalk)	−40 dB	[9]

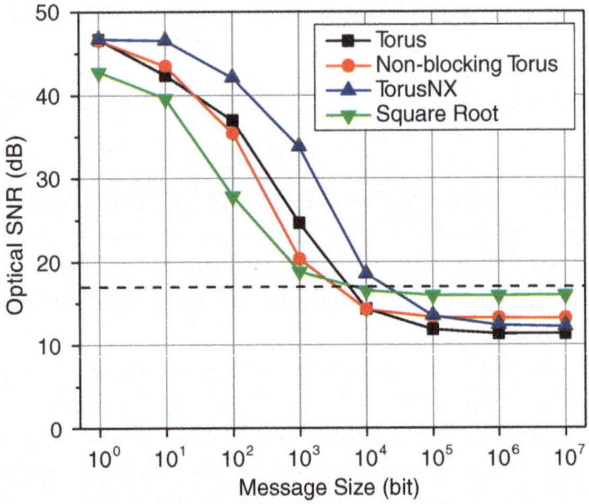

Fig. 5.29 Optical SNR performance for varying message sizes assuming saturated network load, measured at the photodetectors. The line at OSNR = 16.9 dB is where a bit-error-rate of 10^{-12} can be achieved, assuming an ideal binary receiver circuit and orthogonal signaling [11]

the ratio of photonic to electronic activity increases and is reflected by increased optical crosstalk and lower OSNR. This analysis considers when the network is saturated with uniform random traffic. Each network assumes an 8×8 topology.

For short messages, the message transmissions are dominated by the electronic control messages. Therefore optical transmission is less frequent and crosstalk is less likely. In this limiting case, the OSNR is limited by the laser intensity noise. By solving for Eq. 4.10 with the assumed parameters, an OSNR of about 47 dB is achieved, which corresponds well to the simulation results.

For large messages, the latency overhead of the electronic path-setup protocol is amortized by long data transmissions, and the optical network becomes saturated with the long optical messages. In this case, inter-message crosstalk is likely to occur, causing more significant signal degradation. The Square Root topology performs best for large messages with an OSNR of about 16.0 dB. Torus, Non-blocking Torus, and TorusNX achieve OSNRs of 11.3 dB, 13.2 dB, and 12.2 dB, respectively.

Lack of signal integrity ultimately results in erroneous bits detected. If orthogonal signaling and an ideal optimal binary receiver are assumed, then the BER can be calculated using the following Q function [17]:

$$BER = Q\left(\frac{E_b}{\mathcal{N}}\right) \tag{5.3}$$

where E_b is the energy in each bit, and $\mathcal{N}$ is the power spectral density of the noise. The term inside the radical is equivalent to the SNR of the signal. For a BER of 10^{-12}, the network requires a SNR of 16.9 dB (represented by a horizontal line in Fig. 5.29). This indicates that in the large-message cases, none of the networks are able to achieve this level of signal integrity. The achieved BERs for networks using 10^7-bit messages are 1.14×10^{-4} for the Torus, 2.20×10^{-6} for Non-blocking Torus, 2.36×10^{-5} for TorusNX, and 1.31×10^{-10} for Square Root. The high BERs can be lowered by using smaller messages, or mitigated through the use of a higher network-layer error correction scheme.

5.3.2.5 Power Analysis

The network-level power dissipation is a major component in limiting performance scaling of chip-scale systems. Photonic on-chip networks have been shown to drastically outperform electronic networks in both performance and energy, especially in the case of traffic patterns that require large data transmissions [18]. Simulations were conducted to examine the dissipation of the four photonic networks.

Each network is assumed to use the maximum number of wavelengths allowed for the improved 8 × 8 topology with a 30-dB optical power budget according to the results in Figs. 5.19, 5.20, 5.27, and 5.28. The simulator uses the ORION model [19] for electronic router energy dissipation, which is configured for a 32-nm process with a normal voltage threshold transistor type and a V_{dd} equal to 1.0 V. The electronic components in the network are clocked at 1.0 GHz. All electronic routers use a three-stage pipeline model with an 128-bit buffer on each input port and a flit size of 32 bits. All control messages are 32 bits in size. The routers in the torus-like networks use dimension-ordered routing while Square Root uses a unique routing scheme that is optimized to equally distribute load and reduce propagation distance. All routers are modeled with credit-based flow control.

The simulations assume integrated thermal tuners to manage thermal fluctuations in a chip, which will be strongly dependent on application activity. Thermal tuners integrated at each ring in the network assume approximately 1 μW/° K of power dissipation, while the system is assumed to have a mean temperature deviation of 20°. Modulators assume a dynamic dissipation of 85 fJ for every bit transmitted (bit edges) and an additional 30 μW of static power during periods when a constant signal is transmitted (hold periods). Switches exhibit higher dynamic and static dissipation than the ring modulators, at 375 fJ/bit and 400 μW, respectively, due to larger footprints. Photodetector energy is assumed to be 50 fJ/bit. The photonic power dissipation parameters used in this set of simulations are listed in Table 5.5.

The power performance is reported for each of the four networks, and assumes maximum loading with uniform random traffic on 8 × 8 topologies (Figs. 5.30, 5.31, 5.32, and 5.33). In all four network designs, the electronic buffers, crossbar

Table 5.5 Energy dissipation parameters—photonic circuit-switching analysis

Parameter	Value
Lasers (RIN)	−150 dB/Hz
Modulation (modulation index)	16 dB
PSEs—through port extinction ratio	25 dB
PSEs—drop port extinction ratio	20 dB
Crossings (crosstalk)	−40 dB

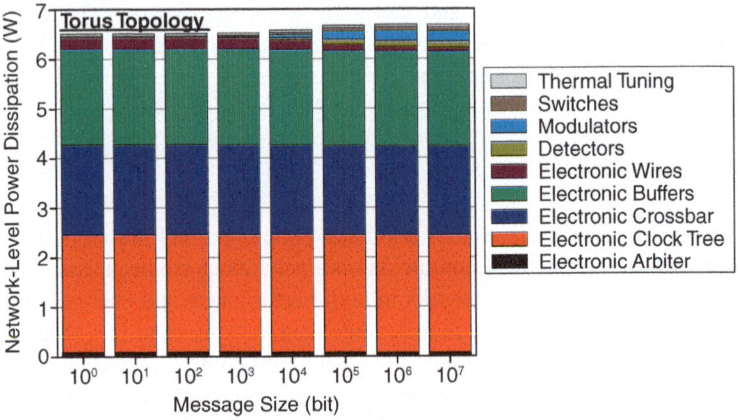

Fig. 5.30 Power-dissipation breakdown of an 8 × 8 Torus topology over varying message sizes [11]

circuit, and clock tree dissipate most of the network power. This is a clear indication that electronic power will remain a significant contributor to total network power dissipation even with photonic integration.

Additional notable trends can be reasoned by relating the power dissipated to the exhibited bandwidth performance of the networks. Figure 5.34 shows the total network performance of the four networks. As the network assumes larger message sizes, the network throughput also rises due to the amortization of the circuit-switching overhead. Congestion of optical traffic on the photonic network plane causes the eventual saturation of the networks. TorusNX achieves the best network bandwidth at 7.80 Tbps, while Square Root, Torus, and Non-blocking Torus obtain throughputs of 3.75 Tbps, 2.45 Tbps, and 669 Gbps, respectively.

Relating back to the four power dissipation figures, as the network achieves higher throughput with larger messages, the ratios in power dissipation shifts from high amounts of wire power dissipation and low photonic device power dissipation to low wire power dissipation and high photonic device power dissipation. This is evidence of the higher photonic network utilization and amortization of the electronic path-setup overhead. Furthermore, the total power dissipated by the electronic components

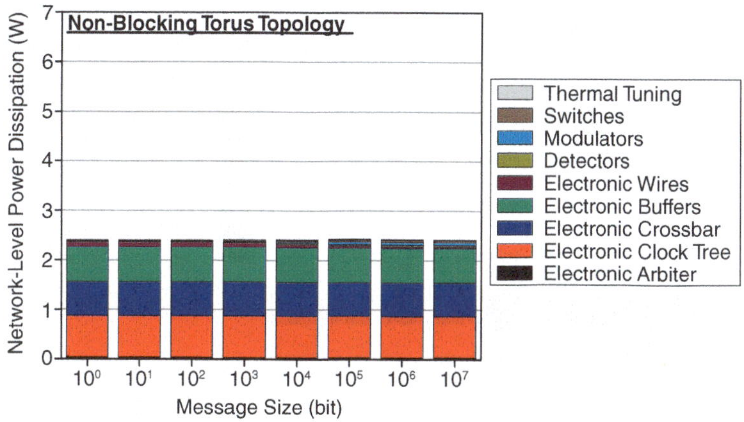

Fig. 5.31 Power-dissipation breakdown of an 8 × 8 Non-blocking Torus topology over varying message sizes [11]

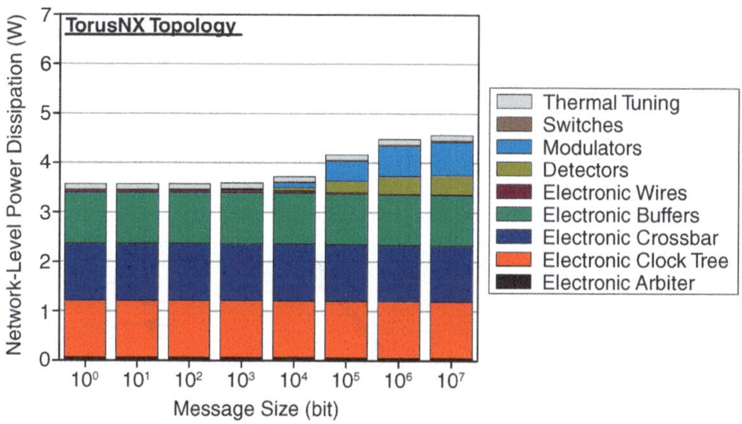

Fig. 5.32 Power-dissipation breakdown of an 8 × 8 TorusNX topology over varying message sizes [11]

in the network remains approximately constant regardless of network throughput since all the data is being sent optically.

Figure 5.35 combines the power and bandwidth results to plot the energy-per-bit efficiency of the networks. For the largest message size, TorusNX and Square Root achieve the best efficiencies at 585 fJ/bit and 681 fJ/bit. Torus achieves an efficiency of 2.73 pJ/bit, while Non-blocking Torus achieves an efficiency of 3.62 pJ/bit. The new network designs attain at least 75 % better efficiency compared to the Torus, and at least 81 % better efficiency than the Non-blocking Torus. This dramatic improvement is attributed to the lower-loss network designs which enable better bandwidth utilization and reductions in the number of required switches.

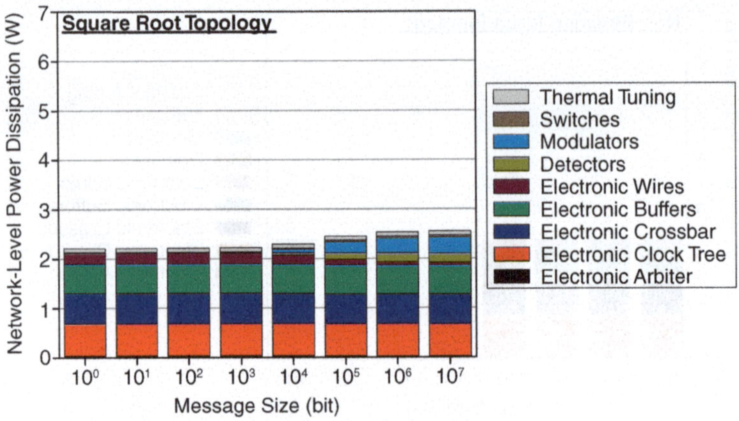

Fig. 5.33 Power-dissipation breakdown of an 8 × 8 Square Root topology over varying message sizes [11]

Fig. 5.34 Total network bandwidth of each network at saturation [11]

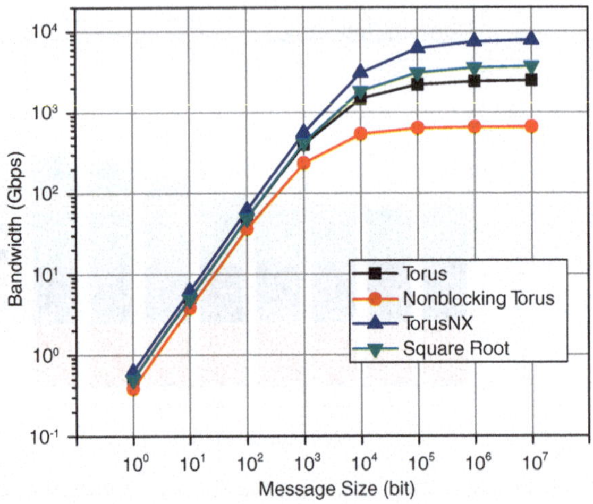

Although the Non-blocking Torus produces a comparatively reasonable absolute power dissipation, the efficiency for larger message sizes is the worst of the four networks. The Non-blocking Torus has the advantage of being non-blocking, however it supports for fewer access points in comparison to the other three network designs results in a dramatic degradation in performance. Note that each network assumes the same topology size. However, the Non-blocking Torus only uses 16 nodes due to the layout constraints. While it may seem reasonable to assume a 32 × 32 Non-blocking Torus so that each network can be normalized to the number of gateways, Fig. 5.20 shows that a 64-gateway version is not possible. The insertion loss penalties cancel the benefits of the non-blocking property, resulting in bandwidth degradation.

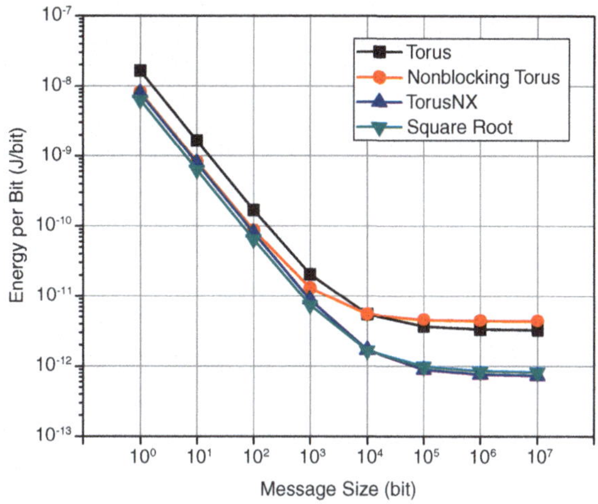

Fig. 5.35 Energy per bit dissipation of each network at saturation [11]

While from an efficiency standpoint, larger message transmissions clearly perform better, the prior crosstalk simulations indicate that the OSNR also decreases with increased message size. This indicates that in order to maintain the high energy efficiency that these photonic topologies can provide, a scheme must be in place to either correct or mitigate these errors.

5.3.3 Effect of Modulation Rate

The relationship between power penalty and data rate of modulator technology presents a unique optimization opportunity when considering how to best incorporate silicon photonic technology. The discussion in Sect. 3.4.1.1 and Fig. 3.6 showed the power penalty associated with replacing a commercially available Mach-Zehnder modulator with a silicon microring modulator for various datarates.

One important aspect of ring modulators is that in practice, as modulation rate increases, the effective power penalty does as well. This effectively means that with an increase in modulation rate comes a higher network worst-case insertion loss, and a decrease in the number of wavelength channels that can be used for a given optical power budget.

Network throughput for the measured power penalty characteristics of the silicon modulator is presented here. This analysis also shows how the throughput could be improved through hypothetical improvements to the modulator. These hypothetical scenarios serve the purpose of motivating device engineers by translating physical-layer metrics into system-level performance metrics. Three cases are examined: the original measured silicon modulator power penalty performance (Fig. 3.6), a hypo-

Table 5.6 Insertion loss parameters—modulation rate study

Parameter	Value
Propagation loss (silicon)	1.5 dB/cm
Waveguide crossing	0.05 dB
Waveguide bend	0.005 dB/90°
Drop into a ring	0.5 dB
Pass by a ring	0.005 dB

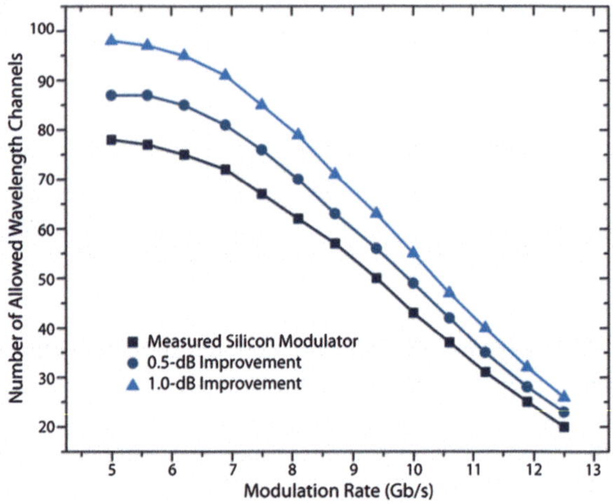

Fig. 5.36 Number of allowed wavelength channels for a wavelength-parallel data signal in a 16 × 16 TorusNX topology as a function of the modulation rate of the silicon microring resonator electro-optic modulator [20]

thetical 0.5-dB power penalty improvement over the original, across all data rates, and a hypothetical 1.0-dB power penalty improvement over the original.

To simulate the network performance, the maximum number of wavelength channels allowed in the network is determined. Eq. 4.9 is adapted by adding a power penalty term:

$$P - S \geq IL_{max} + 10log_{10}n + PP_{mod} \tag{5.4}$$

PP_{mod} is used to account for the power penalty variations of the silicon modulator. Notice that PP_{mod} is unitless which conforms with the formulation of the equation. This effectively normalizes each network configuration to the performance of the modulator, critical for the total system performance.

Using the PhoenixSim simulation environment, the worst-case insertion loss of the 16 × 16 TorusNX topology is determined to be 19.93 dB, with assumed insertion loss parameters in Table 5.6. A 40-dB optical power budget is assumed. Given the relationship between the modulation rate and power penalty in Fig. 3.6, a plot can be created which shows the number of wavelength channels allowed by the network as a function of the modulation rate (Fig. 5.36). The results confirm the large improvement in wavelength parallelism as the physical qualities of the modulator are improved.

With the calculated wavelength utilization and assumed modulation rate, the network-level throughput that will be exhibited by the 16 × 16 TorusNX topology can

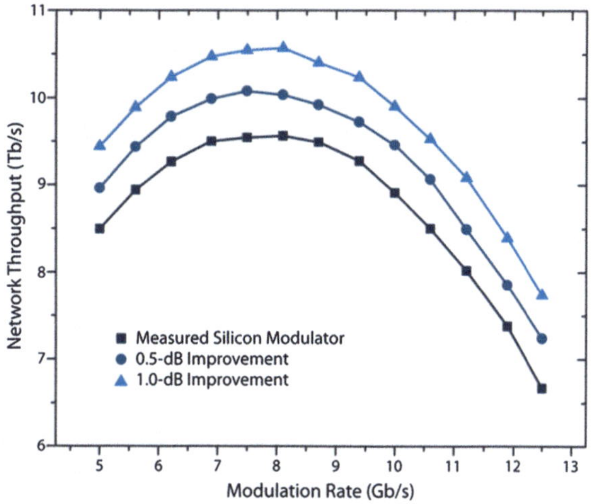

Fig. 5.37 Network throughput of a 16 × 16 TorusNX, including variations in power penalty, as a function of the modulation rate of the silicon microring resonator electro-optic modulator [20]

be determined (Fig. 5.37). Within the simulated parameter range, each set of modulator parameters exhibits a performance peak at approximately the same modulation rate. The total network throughput increases with device improvement, with peaks at 9.6 Tb/s for the original case, 10.1 Tb/s for the 0.5-dB improvement, and 10.6 Tb/s for the 1.0-dB improvement. This trend translates closely to a 0.5-Tb/s improvement in network performance for each 0.5 dB in power penalty improvement. Furthermore, the performance of each modulator case peaks at about the same modulation rate, between approximately 7.5 and 8 Gb/s. Beyond 8 Gb/s, the higher power penalties supersede the advantages gained from faster modulation rates, causing a degradation in network-level performance.

5.4 System Design Considerations

This chapter has thus far explored the system-level effects of physical-layer properties of chip-scale photonic networks. The performance metrics depend largely on the spatial arrangement of photonic components and on physical properties of the devices and network.

This section describes gateway concentration and selective transmission. These design choices do not directly impact the behavior or structure of the photonic network, but can have a dramatic effect on performance. These design choices are applied to the NIF, which is a point in the architecture before data is converted into the optical domain. Later, in Sects. 5.5 and 5.6, simulation results showing the performance impact of gateway concentration and selective transmission will be presented.

5.4.1 Gateway Concentration

Gateway concentration refers to associating more than one network *access point* with a single network *gateway*. Here, the terminology is defined as follows: an *access point* is an entity which requests and receives network communications, such as a core in a CMP. A network *gateway* is an entry point into the photonic network. For direct networks such as a mesh, every network router or node has a gateway associated with it. Gateway concentration is employed to essentially increase the usefulness of the network by ensuring the bandwidth of every network gateway is fully utilized. The risk of using gateway concentration is creating a traffic bottleneck by overwhelming the gateway with more traffic than it can handle. However, the advantage is that access points with less demanding loads can be joined together to reduce the size, footprint, and complexity of the photonic network. This concentration improves the utilization of network resources as well as reduces the physical-layer demands of the photonic components.

Kumar et al. [21] investigated the two different kinds of concentration for electronic NoCs: *integrated* and *external*. These concepts are illustrated in Fig. 5.38 for a four-node segment of a mesh topology. For integrated concentration, in Fig. 5.38b, each access point is given its own NIF, increasing the radix of the network switch from a 5×5 to an 8×8. External concentration, in Fig. 5.38c uses a separate gateway switch to arbitrate access to the main network NIF. External concentration is preferable in several circumstances: (1) when increasing the radix of the network switch is especially complex or costly, (2) when the expected global communication requirements of individual access points is sufficiently low such that sharing a single network gateway will not significantly affect performance, and (3) when it is desirable to use different network technologies or architectures for local and global communications.

Integrated and external gateway concentration for photonic networks dictates whether photonic or electronic network resources need to be adjusted since the electronic-photonic boundary is at the NIF. Integrated gateway concentration indicates a change to the design of the photonic gateway switch. External gateway concentration changes are entirely electronic, since it inserts an electronic switch within the NIF before conversion from electronics to photonics.

Integrated gateway concentration increases the complexity of the photonic network design and therefore increases the network-level insertion loss. As shown earlier, increases to insertion loss can be detrimental to photonic network scalability. For this reason, a good design practice is to use *external* concentration for circuit-switching.

The use of external gateway concentration has several significant implications on power and performance. First, it is generally understood that electronics is more efficient for transmission of data across short distances on a chip because of the maturity of CMOS technology and design, which makes it a natural choice to implement the gateway switch with electronics. However, to avoid having to provide sufficient buffering at the NIF to hold entire (potentially large) photonic transmissions, the

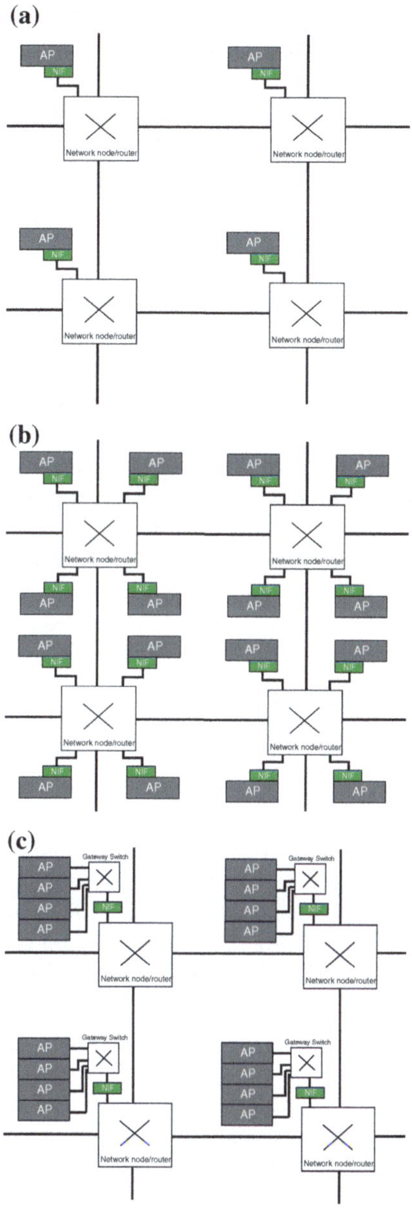

Fig. 5.38 **a** Four unconcentrated network gateways each with a single access point. **b** Four network gateways with 4-way integrated concentration. Each gateway connects to four separate access points. **c** Four network gateways with 4-way external concentration. The gateway is the same as in the normal case, but connects to an auxiliary switch to connect to the access points. Each gateway connects to a switch, which subsequently is connect to four access points

path setup protocol should extend into the gateway switch domain so that paths are reserved *through* the gateway switch both from the sending and receiving side. This allows data to remain in the memory space of the processor as opposed to the NIF. The important implication of this idea is that the gateway switch must be able to bandwidth-match the photonics. This may require a high-speed electronic crossbar which would have a high cost overhead in power. If concentration must be used to increase the scalability of the network, a potential compromise to balance power utilization is to lower modulation rates to avoid costly SerDes technology.

5.4.2 Selective Transmission

Fundamentally, packet-switched networks are suitable for handling small messages with random communication patterns while circuit-switching is best suited for large data streams with static communication patterns. From the perspective of electronic and photonic networks, an electronic packet-switched network would be much more efficient at transporting small packets over small distances, while circuit-switching would be more useful for larger messages with relatively static communication patterns. Photonic networks can potentially take advantage of this performance difference by designing the NIF to filter traffic into these two classes. Photonic circuit-switched networks by design already possess a light-weight packet-switched electronic network for control and arbitration. The modified NIF can take advantage of this electronic network for transmitting smaller messages. This policy is called *selective transmission* and was first proposed for photonic circuit-switched networks by Hendry et al. [18].

The chief consideration in determining the operation of selective transmission is the size threshold for determining transmission on the control network or the photonic circuit-switched network. The size threshold has an impact on the required performance of the electronic control network as well as the final power and throughput performance. Determining an optical size threshold relies on the consideration of a number of factors, such as topology, expected application behavior, and size requirements of the electronic control network.

Figure 5.39a illustrates a communication pattern where a large number of small messages are used to signal control information, (e.g. coherence and barrier messages), and larger data messages are widely distributed in size. A natural choice of selective transmission policy would be to set a size threshold such that all small control messages are sent on the electronic network, while all larger data messages are sent optically. A system architect will need to determine the required electronic buffer size and channel width such that power and performance requirements are met. The same consideration is also applicable to other scenarios such as the examples in Fig. 5.39b, c. This optimal size threshold is not immediately apparent and will require a thorough exploration of the costs and benefits.

An alternative selection policy would leverage spatial locality of access points. Messages which are destined for nearby access points would utilize the control network, while long-distance communications would use the photonic network. This has the benefit of reducing multi-hop communications which are costly in terms of power dissipation and electronic network congestion.

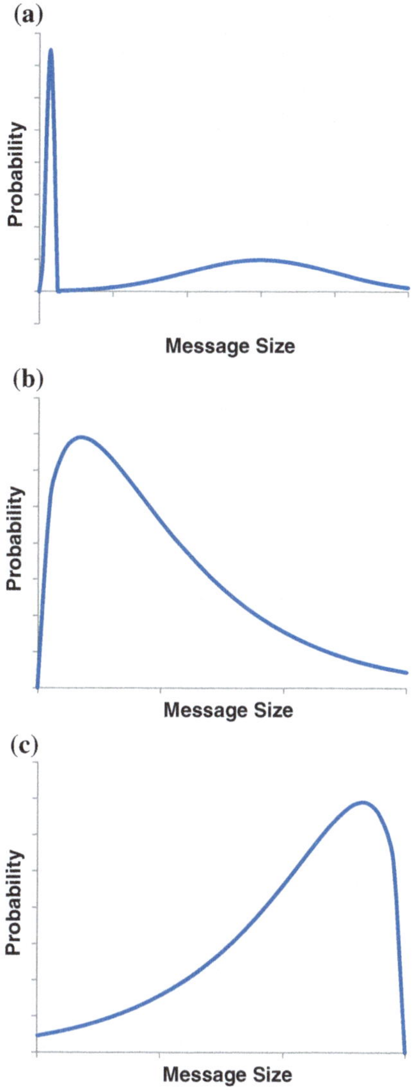

Fig. 5.39 Example message size distributions for hypothetical applications. **a** Application containing many small control messages and a large size variance of larger messages, **b** application that contains a distribution of message sizes weighted towards smaller sizes, and **c** application that contains a distribution of message sizes weighted towards larger sizes

5.5 Evaluating Photonic Circuit Switching with Scientific Applications

Consideration of real application traffic can be beneficial to the understanding of true network performance since real applications exhibit behavior that is dramatically different in comparison to synthetic traffic. This section investigates different networks using message-passing interface (MPI) based scientific applications.

5.5.1 Application Description

This analysis considers four high-performance computing applications: Cactus, GTC, PARATEC, and MadBench.

The first evaluated application is *Cactus* [22], an astrophysics computational toolkit designed to solve coupled nonlinear hyperbolic and elliptic equations that arise from Einstein's Theory of General Relativity. Consisting of thousands of terms when fully expanded, these partial differential equations (PDEs) are solved using finite differences on a block domain-decomposed regular grid distributed over the processors. The Cactus communication characteristics reflect the requirements of a broad variety of PDE solvers on non-adaptive block-structured grids.

The Gyrokinetic Toroidal Code (*GTC*) is a 3D particle-in-cell (PIC) application developed to study turbulent transport in magnetic confinement fusion [23]. GTC solves the non-linear gyrophase-averaged Vlasov-Poisson equations in a geometry characteristic of toroidal fusion devices. By using the particle-in-cell method, the non-linear PDE describing particle motion becomes a simple set of ordinary differential equations (ODEs) that can be solved in the Lagrangian coordinates. GTC's Poisson solver is localized to individual processors, allowing the communication traces to only reflect the needs of the PIC core.

The PARAllel Total Energy Code [24] (*PARATEC*) is a materials science application that is widely used to study properties such as strength, cohesion, growth, and transport for materials like nanostructures, complex surfaces, and doped semiconductors using the Density Functional Theory (DFT) method. In solving the Kohn-Sham equations using a plane wave basis, part of the calculation is carried out in real space and the remainder in Fourier space using specialized parallel 3D FFTs. The all-to-all communication used to implement the 3D data transpose for the FFT is the most demanding portion of PARATEC's communication characteristics.

The fourth application examined is *MADbench* [25], a benchmark based on the MADspec cosmology code. MADspec calculates the maximum likelihood angular power spectrum of the cosmic microwave background (CMB). MADbench tests the overall performance of the subsystems of real parallel architectures by retaining the communication and computational complexity of MADspec and integrating a dataset generator that ensures realistic input data. Much of the computational load of this application is due to its use of dense linear algebra, which is reflective of the requirements of a broader array of dense linear algebra codes in scientific workloads.

Together, these four applications represent a broad subset of scientific codes with particular communication requirements both in terms of communication topology and volume of communication. For example, the nearest-neighbor Cactus communication represents components from a number of applications characterized by stencil-type behavior. Results are therefore generally applicable to a broad range of numerical computations.

Figure 5.40 shows the spy plots of the eight benchmarks in this study. Each application utilizes 64 independent processes on 64 compute nodes. The spy plots illustrate the communication volume between each pair of processes: a white square

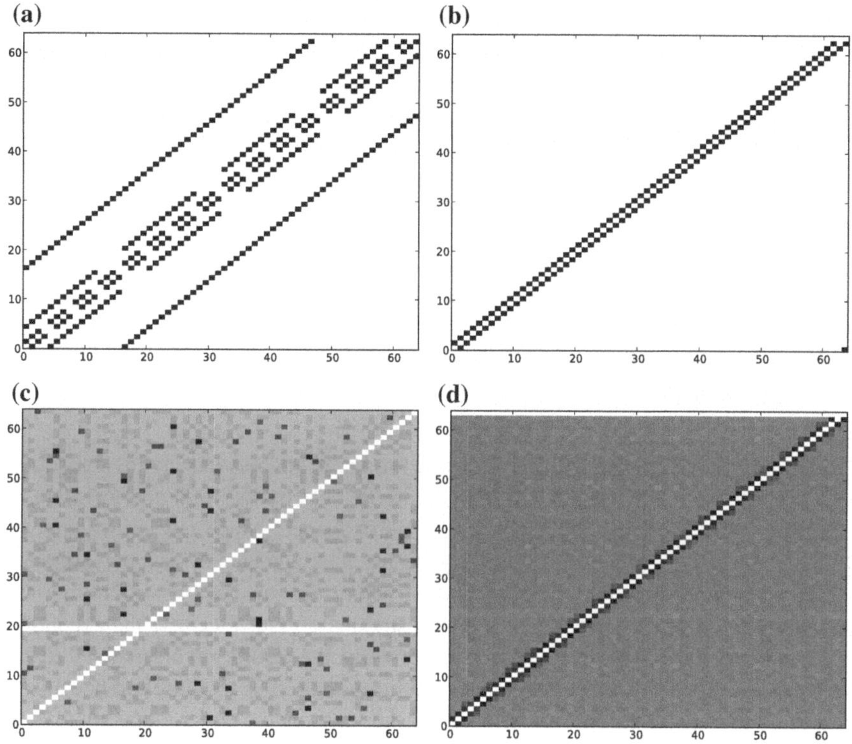

Fig. 5.40 Spyplots for the studied applications [18]

Table 5.7 Application statistics

Benchmark	Num. phases	Num. messages	Total size (B)	Avg. message size (B)
Cactus	2	285	7296000	25600
GTC	2	63	8177148	129796
MADbench	195	15414	86516544	5613
PARATEC	34	126059	5457332	43.3

at the coordinate (p_i, p_j) in the plot represents no communication, while darker shades of gray represent increasing volumes of communication between the pair. Quantitative details of the different benchmarks are given in Table 5.7.

5.5.2 Studied Network Architectures

The network architectures described in this section will be evaluated using traces from the applications described in Sect. 5.5.1. Networks include both conventional electronic networks and photonic networks. This evaluation assumes a CMP with 64 processors arranged in an 8×8 grid arrangement.

The modeled electronic networks include the mesh, concentrated mesh, and concentrated torus topologies (illustrated in Fig. 5.41). The mesh topology is used as a baseline comparison. In contrast with other exotic electronic network topologies, the mesh is simple to implement due to its use of relatively low radix switches in a regular 2D planar layout. This study also incorporates the gateway concentrated electronic networks, as discussed in Sect. 5.4.1.

The photonic network used in this study is the folded torus topology shown in Fig. 5.14. A selective transmission variant with a size threshold of 256 B is also included.

5.5.3 Evaluation

The reported metrics are as follows: (1) performance is analyzed via the execution time of the benchmark or application, (2) energy cost by the total energy spent in execution, and (3) energy efficiency by the performance gained per unit of energy. Note that while typical network comparisons use message latency as a performance metric, such analysis would underscore the true performance of the system by only examining the transmission speed of single streams of data. Because the execution times and energies of the benchmarks vary broadly, the results are normalized to the electronic mesh performance. An electronic mesh is used as the baseline because it represents the most straightforward engineering approach to interconnecting cores for emerging CMP designs.

The scientific application experiments are conducted using fifty random mappings of processes to compute nodes to develop a statistical view of the networks responsiveness to varying communication mappings. This random mapping technique is used since the optimal mapping is not known. Application results are therefore shown using the average performance, with error bars indicating minimum and maximum behaviors.

5.5.3.1 Execution Time

Figure 5.42 presents the application execution time speedup achieved by the examined NoC architectures relative to the execution time of the baseline electronic mesh. Values start at one, which indicates even performance with the baseline. The concentrated photonic networks clearly outperform the other interconnect configurations for both Cactus and GTC, which is expected from the large-message characteristics. The photonic networks do not perform as well for the MADBench and PARATEC applications primarily because those benchmarks exhibit all-to-one and broadcast communication patterns, which are expected to behave poorly in circuit-switched networks. For these types of applications, wavelength-routed inter-core networks would likely be more appropriate. In addition, these two benchmarks use significantly smaller message sizes (see Table 5.7). The selective transmission networks

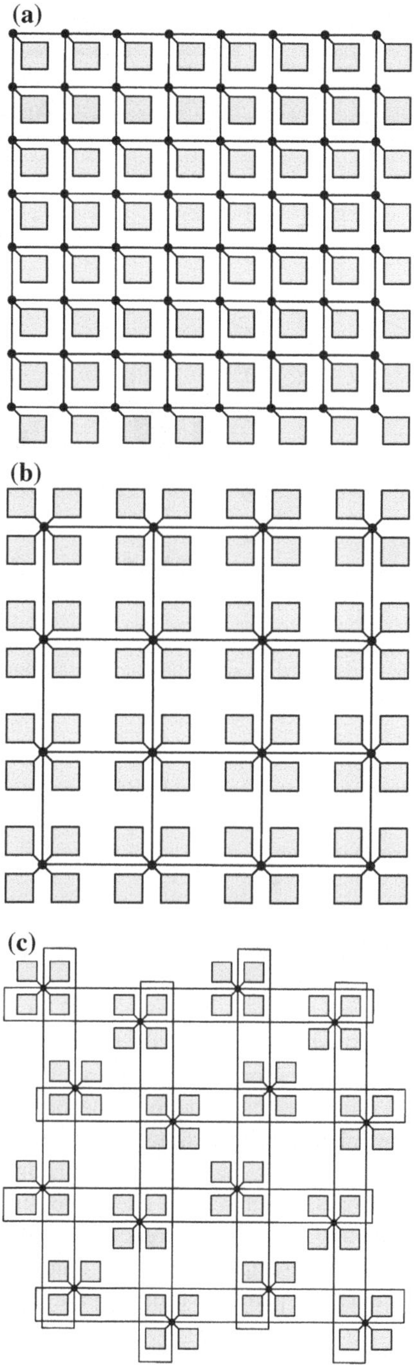

Fig. 5.41 Mesh, concentrated mesh, and concentrated torus topology. The concentrated topologies require a larger-radix switch, but reduce the average hop count [18]

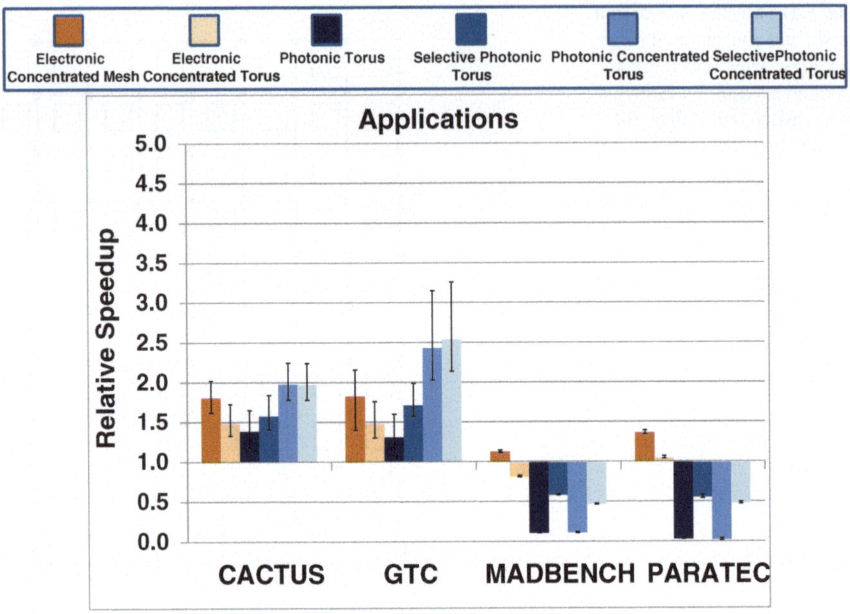

Fig. 5.42 Average application execution time performance for various photonic and electronic networks based on simulation of fifty random process-to-node mappings. Results are normalized to the electronic mesh. Error bars indicate minimum and maximum achieved performance [18]

narrow the performance difference somewhat, but still do not achieve the nominal performance of the electronic mesh network, similar to the synthetic traces using small messages.

5.5.3.2 Energy Consumption

Figure 5.43 presents results based on total energy consumption; the plot shows the inverse of consumption (i.e. the energy savings) relative to the electronic mesh baseline. The photonic networks are clear winners for Cactus and GTC, showing over 10× improvement due to the decoupling of distance, bandwidth, and power during optical transmission. Since the circuit-switched photonic network does not consume power per-hop, the energy usage is much lower than the packet-switched electrical networks, which require energy consumption in order to make routing decisions at each hop. Because photonics is completely decoupled from distance travelled with respect to energy spent during transmission, it provides higher benefits when communication pairs are further apart. The MADBENCH and PARATEC applications

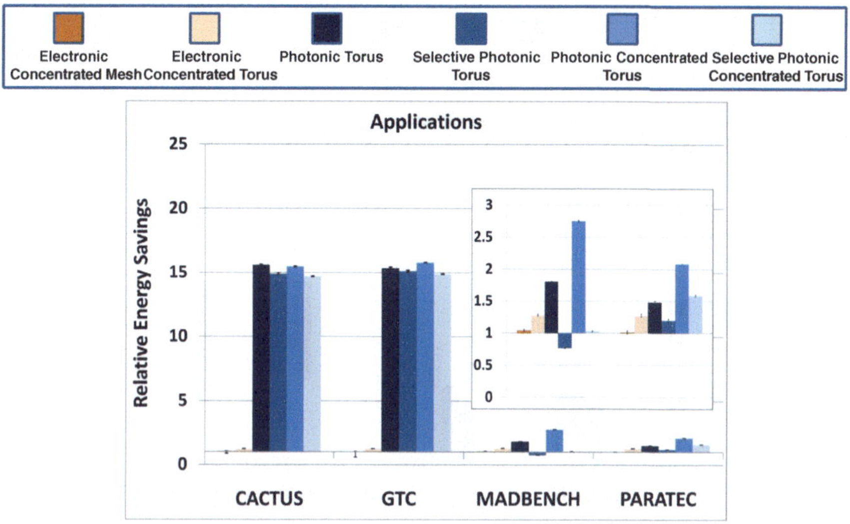

Fig. 5.43 Average application energy savings (inverse of energy consumption) for various photonic and electronic networks based on simulation of fifty random process-to-node mappings. Results are normalized to the electronic mesh. Error bars indicate minimum and maximum achieved performance. The inset plot provides a magnified view of the MADBENCH and PARATEC results for clarity [18]

do consume more energy when selective transmission is turned on, because most of the messages traverse the electrical network.

5.5.3.3 Network Efficiency

The final metric considered is the performance gained for every unit of energy spent, which is effectively a measure of network efficiency. This metric is calculated by multiplying the network execution time by the energy consumed. Figure 5.44 shows these results normalized to the electronic mesh. The plot is oriented in a way such that values greater than unity indicate a better performance per energy than the electronic mesh.

The applications with small messages perform poorly on photonic networks. Although network speedup is reasonable for some photonic networks in Fig. 5.42, and energy gains are achieved for some photonic networks in Fig. 5.43, the overall network performance is not improved over the electronic mesh when message sizes are small.

However, the photonic networks' energy efficiency improvement over the electronic mesh for traces with large message sizes is amplified by the gains in both speedup and energy, resulting in improvements of over 20×.

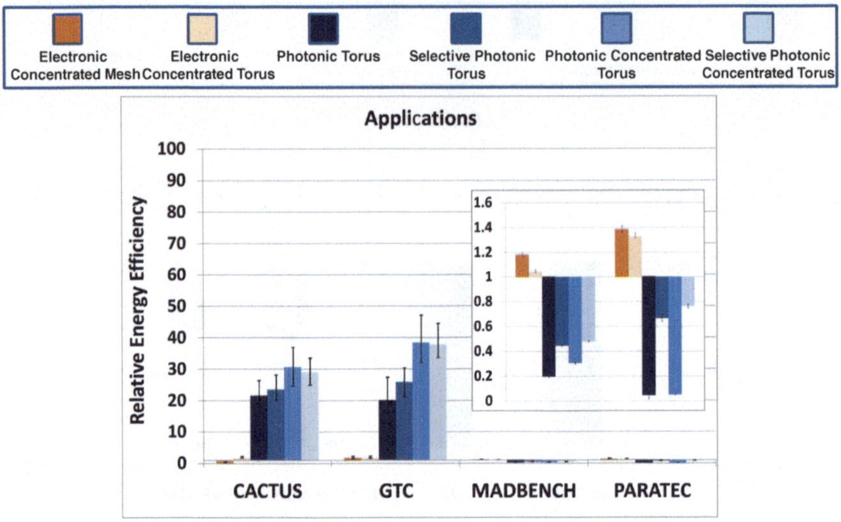

Fig. 5.44 Average application energy efficiency for various photonic and electronic networks based on simulation of fifty random process-to-node mappings. Results are normalized to the electronic mesh. Error bars indicate minimum and maximum achieved performance. The inset plot provides a magnified view of the MADBENCH and PARATEC results for clarity [18]

5.6 Off-chip Memory Access

One of the most important subsystems in a computer is the off-chip memory, and how data is moved to, from, and around the chip. Photonics has some unique potentials for IO communication because of the nature of the transport medium. Since waveguides are bit-rate transparent through the chip edge, the same modulators and detectors can be used for both on-chip and off-chip communication. Furthermore, distance is much less of a factor in terms of bit rate and power when considering distances at the board-scale. Finally, WDM offers extremely high bandwidth density for links crossing the chip boundary, which is an enormous advantage considering the limitations imposed by chip package electronic pin counts. A description of how circuit-switched networks can accomplish core-memory communication is provided next. Hendry et al. [2] first proposed this by extending the same circuit-switched network resources and path-setup protocol out to the chip edge for DRAM reads and writes.

Figure 5.45 shows one possible implementation of circuit-switched memory access using lateral coupling to optically connect memory to a processor chip. A memory controller interacts with the electronic control network to establish end-to-end optical circuits between access points in the network and individual circuit-accessed memory modules (CAMMs). The memory controller issues row/column access commands to the DRAM chips using dedicated modulators and wavelengths. CAMMs multiplex and demultiplex the wavelength channels of a WDM signal to and

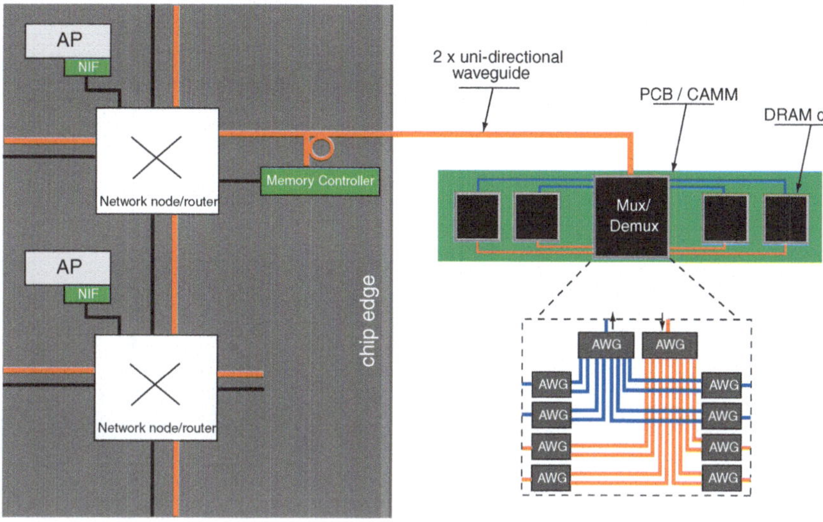

Fig. 5.45 Structure of the memory controller to circuit-accessed memory module (CAMM) link. The on-chip memory controller transmits optical memory accesses via dedicated modulators and waveguides. The CAMM is composed of a optical multiplexer and demultiplexer attached to several DRAM chips with integrated optical transceivers [2]

from several DRAM chips using arrayed waveguide gratings (AWG). This aggregation allows read and write access to all DRAM chips in parallel.

Figure 5.46 shows the control flowchart for the memory controller and how it interfaces the electronic path setup protocol with DRAM access commands. An important differentiator between traditional memory access and the access pattern imposed by the circuit-switching path-setup protocol is the number of concurrent memory requests. Traditional memory subsystems can process and re-order multiple requests for better memory access efficiency.

CAMM access simplifies memory by recognizing that circuit switching can only provide a single data link at a time. Memory read requests are sent to the memory control via the electronic control network without allocating any optical resources. If a read request can be processed, then the circuit switching policy is employed with the memory controller as the source and the requesting access point as the destination. Write requests reserve optical resources as they move towards the memory controller. The controller is responsible for transmitting a path-ACK message in order to notify the source access point that the memory and optical path is ready for transmitting data. If the memory or the network is not ready, then a path-blocked message is returned in order to free up the network resources. Like circuit-switching itself, this design is best suited for applications with large data sets and streaming transmissions.

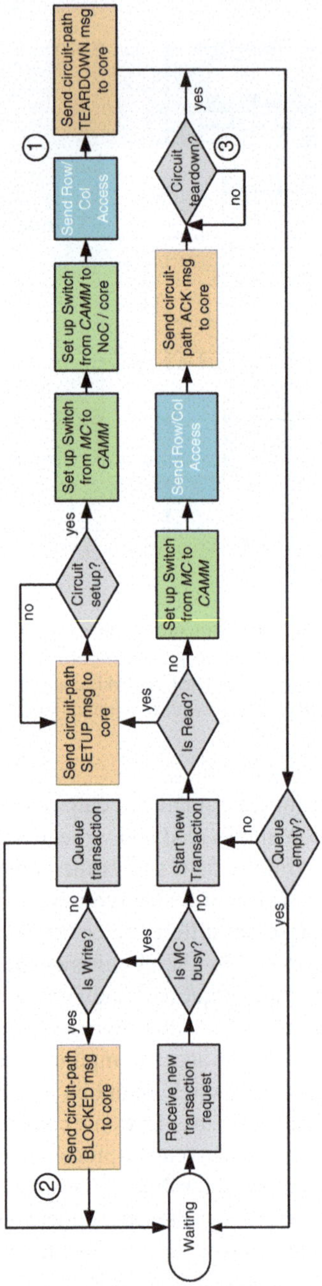

Fig. 5.46 Control flowchart of circuit-switched memory controller, interacting with electronic control network [2]

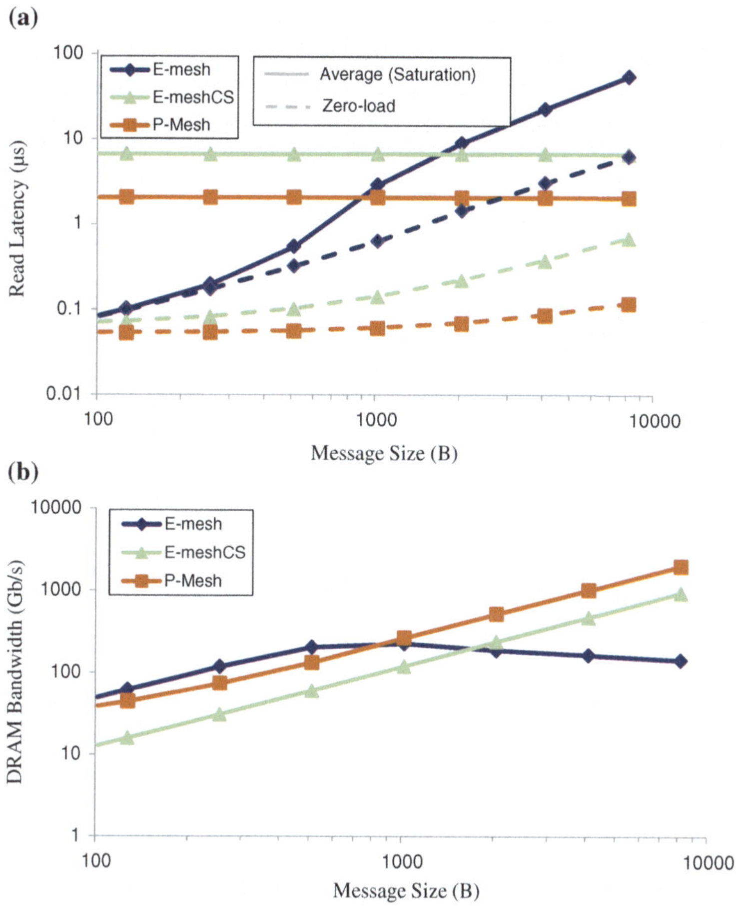

Fig. 5.47 Performance characteristics of DRAM subsystem for packet-switched E-Mesh, circuit-switched E-Mesh, and photonic circuit-switched P-Mesh [2]

This next section characterizes through simulation the performance of the aforementioned circuit-accessed memory subsystem design. This analysis assesses the effectiveness of circuit-switched memory access and photonic technology. A comparison is performed between an electronic packet-switched mesh (E-Mesh), electronic circuit-switched mesh (E-MeshCS), and a photonic circuit switching mesh (P-Mesh). Each network is loaded with random DRAM accesses with equal probability of reads and writes. Each network is 8×8 nodes in size, with a memory controller located at each node on the perimeter of the chip. The important metrics here are read latency for reducing CPU stall time, and total DRAM bandwidth.

Figure 5.47a shows the zero-load read latency and average latency under saturation. The simulation model is based on the assumption that each memory access message is seeking a single contiguous portion of memory. This is a reasonable assumption for many systems (e.g. embedded systems reading sensor data, stream-processing systems, or large cache/buffer lines). The circuit-switched networks (both electronic and photonic) are able to achieve significantly lower zero-load read latency, becoming smaller with larger messages. This is because the memory controller is not trying to optimize DRAM utilization by scheduling small accesses, but reads whole DRAM array rows at a time. Under contention, the packet-switched E-Mesh has comparatively lower latency for smaller messages because of the path-setup overhead. This overhead becomes amortized in the circuit-switched networks for larger messages. The electronic circuit-switched network uses the same transmission strategy as P-Mesh, but it experiences significantly higher latencies as message sizes increase. This is a result of the pin-count limit imposed on electronic I/O. In contrast, photonics can leverage more bandwidth with WDM. Figure 5.47b shows the total DRAM bandwidth achieved. The circuit-switched networks scale with message size, while the packet-switched performance drops off at larger messages due to packetized and saturate the DRAM subsystem.

5.7 Evaluating Photonic Memory Access Using Embedded Applications

This section describes the evaluation of the memory network architecture using the application modeling framework called *Mapping and Optimization Runtime Environment* (MORE). MORE was used to collect traces from the execution of high-performance embedded signal and image processing applications.

The MORE system, based on pMapper [26], is designed to project a user program written in Matlab onto a distributed or parallel architecture and provide performance results and analysis. The MORE framework translates application code into a *dependency-based instruction trace*, which captures the individual operations performed as well as their interdependencies.

5.7.1 Embedded Applications

The following small embedded mini-applications were traced with MORE to be able to run them back over a hypothetical photonic network design.

Projective Transform. When registering multiple images taken from various aerial surveillance platforms, it is frequently advantageous to change the perspective of these images so that they are all registered from a common angle and orientation (typically straight down with north being at the top of the image). In order to do this, a process known as *projective transform* is used [27].

Projective transform takes as input a two-dimensional image M as well as a transformation matrix t that expresses the transformational component between the angle

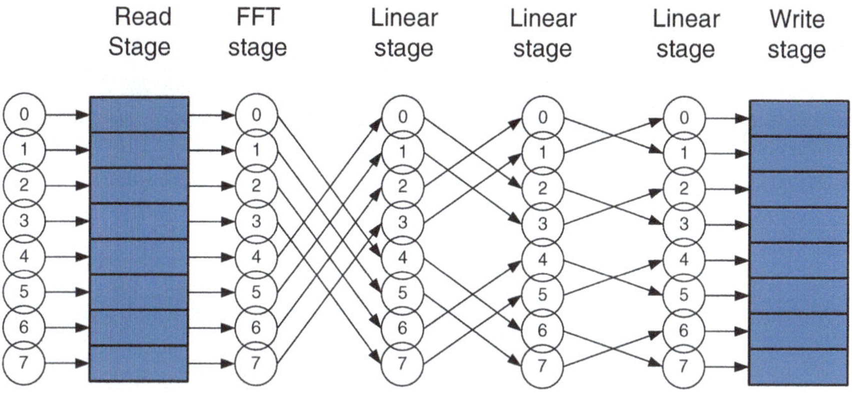

Fig. 5.48 Communication pattern of the Cooley-Tukey algorithm for computing the FFT. Each labeled number indicates a processing node [2]

and orientation of the image presented and the desired image. The projective transform algorithm outputs M', or the image M after projection through t. To populate a pixel p' in M', its x and y positions are back-projected through t to get their relative position in M, p. This position likely does not fall directly on a pixel in M, but rather somewhere between a set of four pixels. Using the distance from p to each of its corners as well as the corner values themselves, the value for p' can be obtained.

MORE allows the retention of identical image and projections sizes while still inducing data movement in the projection process as well as investigating various transformation matrices. This experiment considers this application on various image sizes where the image orientation is rotated by 90°.

Matrix Multiply. Matrix multiplication is a common operation in signal and image processing, where it can be used in filtering as well as to control hue, saturation and contrast in an image. It is a natural candidate for consideration on our architecture, given that multiple data points need to be accessed and then summed to form a single entry in the result.

While various algorithms for matrix multiplication can be considered for matrices of any dimension, this analysis will focus on an inner product algorithm over square matrices. Here, in an $N \times N$ matrix, each entry is generated by first multiplying together two vectors of size N (corresponding to a row and a column), and then summing the entries in the resulting vector to form a single entry in the result.

The inner product algorithm requires time proportional to N^3. While the best known algorithm for matrix multiply is $O(N^{2.376})$, the constants in the algorithm make it infeasible for all but the largest of matrices. Even Strassen's algorithm [28], with a bound of $O(N^{2.806})$ is frequently considered too cumbersome and awkward to implement, especially in a parallel environment. Though more computationally expensive, the inner product algorithm also lends itself more naturally to a parallel implementation, making it our algorithm of choice.

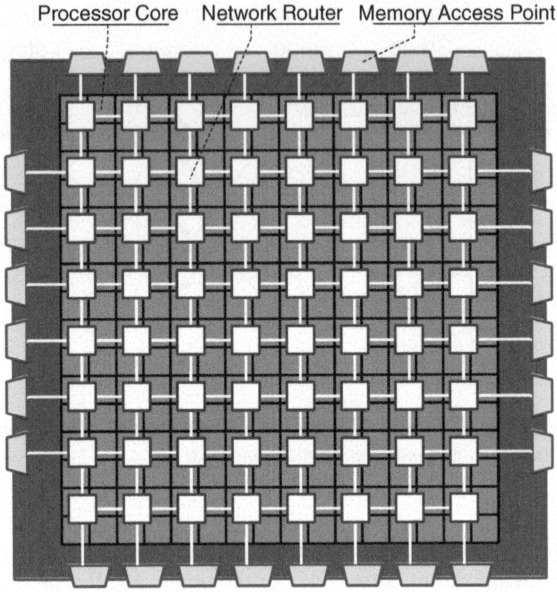

Fig. 5.49 Abstract illustration of 8 × 8 mesh network-on-chip with peripheral memory access points [2]

Fast Fourier Transform. Computing the Fast Fourier Transform (FFT) of a set of data points is an essential algorithm which underlies many signal processing and scientific applications. In addition to the widespread use of the FFT, the inherent data parallelism that can be exploited in its computation makes it a good match for measuring the performance of networks-on-chip. The Cooley-Tukey method is a common algorithm used to compute the FFT in parallel [29]. The communication patterns and computation stages for 8 nodes are shown in Fig. 5.48. The FFT is executed with each core begins with 2^{10}, 2^{12}, 2^{14}, 2^{16}, and 2^{18} samples, and average the results.

5.7.2 Network Architectures

Three different network architectures are considered: Electronic packet-switched (Emesh), Electronic circuit-switched (EmeshCS), and Photonic circuit-switched (PmeshCS). All three use an 8 × 8 mesh topology to connect a grid of 64 network nodes with DRAM access points on the periphery. An abstract illustration of this setup is shown in Fig. 5.49.

The Emesh and EmeshCS use *integrated gateway concentration* [21] of 4 cores per network gateway, for a total core count of 256. Table 5.8 shows some of the simulation parameters used for the electronics.

The photonic circuit-switched mesh uses *external gateway concentration* [21] to simplify the complexity of photonic switch design required at the access point.

Table 5.8 Electronic simulation parameters

Parameter	Emesh	EmeshCS	PmeshCS (PS1)	PmeshCS (PS2)
Chip IO parameters				
Physical I/O per MAP	64	32 (diff pair)	2 (w/128 λ)	2 (w/45 λ)
I/O bit rate	1.6 GT/s	10 Gb/s	2.5 Gb/s	2.5 Gb/s
Bandwidth I/O link (Gb/s)	102	320	320	112
NoC electronic parameters				
Clock frequency (GHz)	1.6	1.0	1.0	1.0
Data plane frequency (GHz)	–	2.5	2.5	2.5
Buffer size (b)	1024	128	128	128
Virtual channels	2	1	1	1
Control plane V_{DD}	0.8	0.8	0.8	0.8
Control plane V_{th}	Norm	High	High	High
Data plane V_{DD}	–	1.0	1.0	1.0
Data plane V_{th}	–	Norm	Norm	Norm
Electronic channel width	64	32 (128 for data)	32	32
Bandwidth On-chip link (Gb/s)	102	320	320	112
DRAM parameters				
Base DRAM frequency (MHz)	1066	1066	1066	1066
Arrays per Bank	8	32	32	16
Chips per DIMM	8	10	10	8
DIMMs per MAP	2	1	1	1
Total memory per MAP	2 GB	2 GB	2 GB	2 GB
Bandwidth per DIMM (Gb/s)	128	320	320	128

Designs of 4×4 photonic switches were explored previously in Sect. 5.2.2, however the mesh router requires five ports (four external directions and one to the processor core). Therefore an alternative design must considered.

Figure 5.50 depicts two designs for the photonic 5-port ring resonator-based broadband data switch used in the circuit-switching router for the PmeshCS, designated as PS-1 and PS-2.

The PS-1 design augments the StraightPath 4×4 Switch design with modulator and detector banks between the waveguide pairs in the south and east direction (similar to the technique used for the TorusNX topology, see Sect. 5.3.2.3). As a result, the switch has a small number of rings and low insertion loss, being able to use 128 wavelengths based on a rough insertion loss analysis. However, it exhibits blocking behavior when certain switch configurations are used (e.g. when the detector bank is being used, the east-bound port is blocked).

The PS-2 switch design is based on a full ring-matrix crossbar switch, removing rings used for U-turns and rerouting waveguides to create default unswitched paths. The PS-2 switch uses more rings and has greater insertion loss, yielding 45 wavelengths, but is fully nonblocking.

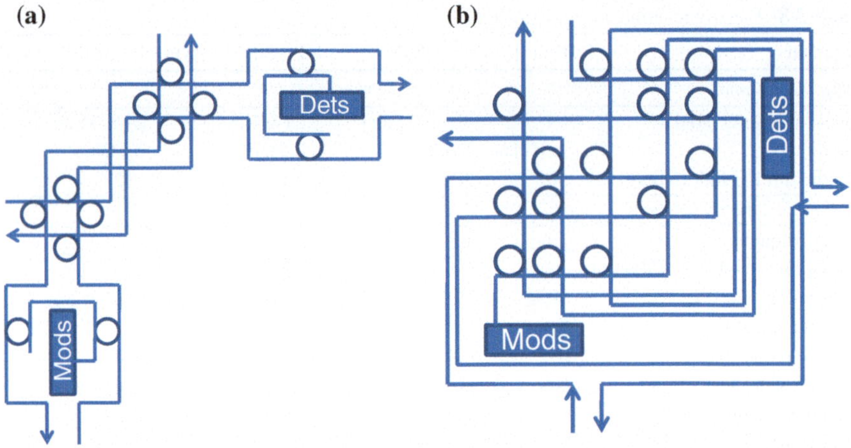

Fig. 5.50 Two designs for a 5-port photonic switch for the PmeshCS [2]

Table 5.9 Results for performance, network power, and improvement over electronic mesh in signal and image processing applications

Network	Projective transform			Matrix multiply			Fast Fourier transform		
	Power (Watts)	Perf. (GOPS)	Impr. (GOPS/W)	Net. Pow. (Watts)	Perf. (GOPS)	Impr. (GOPS/W)	Net. Pow. (Watts)	Perf. (GOPS)	Impr. (GOPS/W)
Emesh	11.2	1.04	1x	11.1	0.78	1x	11.4	1.75	1x
EmeshCS	19.0	47.3	26.9x	15.8	31.82	29.01x	11.2	4.74	2.82x
PS-1	4.37	27.80	68.6x	4.35	26.51	87.64x	4.28	4.32	6.72x
PS-2	2.21	17.76	86.7x	2.17	13.48	89.33x	2.15	3.12	9.67x

5.7.3 Simulation Results

Table 5.9 shows the averaged results for the different network configurations across the three applications described earlier. The plots show network-related power, total system performance, and total system efficiency. Performance is measured in giga-operations per second (GOPS) and system efficiency is measured in GOPS/W. The measurements are normalized to the Emesh for comparison. In all cases, the circuit switched networks achieve considerable improvements in both performance and power over the Emesh.

For the Projective Transform and Matrix Multiply, the EmeshCS consumes some additional power to achieve considerable gains in performance. The photonic networks also perform significantly better than the Emesh, though at much lower power than the EmeshCS. The PS-2 generally consumes less power because it has less modulators (but also less bandwidth), and uses non-blocking switches. This reduces the network blocking frequency and consequentially the power dissipation. The FFT exhibits a different communication and memory access behavior than the other appli-

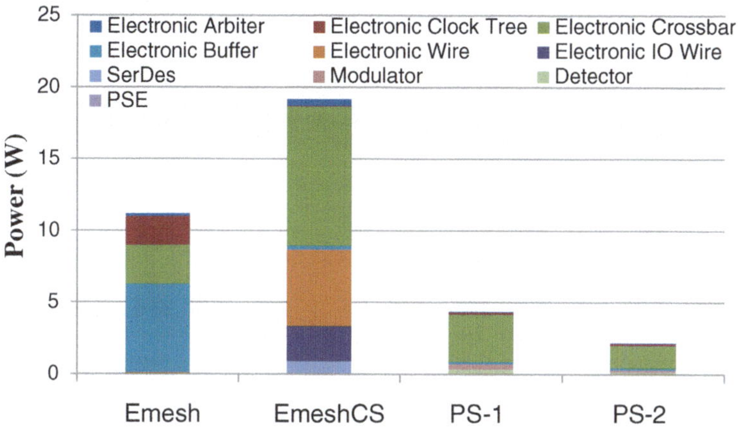

Fig. 5.51 Projective Transform network power breakdown [2]

cations. In the case of the PS-2 PMesh, it is able to achieve almost an order of magnitude improvement over Emesh.

The breakdown of power consumption for the various network components is shown in Fig. 5.51 for the Projective Transform, one of the more network-active applications. The Emesh power dissipation is comprised mostly of buffer, crossbar, and clock tree power. EmeshCS alleviates buffer power as intended, but at the cost of crossbar and wire power in the higher-frequency data plane. Finally, the photonic networks achieve drastically lower power through distance-independent efficient modulation and detection.

5.8 Architectures Using Deposited Multi-layer Devices

Multi-layer devices using deposited materials are important to consider for photonic NoC design for their potential in reducing insertion loss and easing the fabrication process. Though not as versatile as pure crystalline silicon, materials such as silicon nitride (Si_3N_4) and polycrystalline silicon can be combined to not only implement the same functions, but also add extra dimensions of design freedom in photonic layout. The chief benefit of multi-layer devices for photonic circuit switching is the elimination of waveguide crossings which reduces the insertion loss of the network and consequentially greatly improves the scalability of the network. This approach was first investigated by Biberman et al. [30].

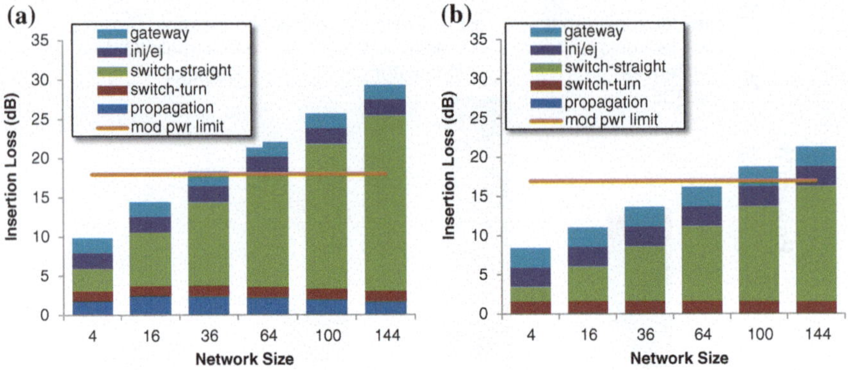

Fig. 5.52 Insertion loss of single- and multi-layer photonic mesh [30]

5.8.1 Multi-Layer Mesh

The insertion loss of a photonic mesh using multi-layer deposited materials is considered here. The StraightPath 4 × 4 Switch in Fig. 5.8c can be created by implementing the East-West paths with one layer of nitride and the North-South paths with another layer of nitride in order to completely eliminate waveguide crossings from the switch. Switches are implemented in polycrystalline silicon layer that is sandwiched between the two nitride layers. Figure 5.52 shows the insertion loss for various size networks with a constant channel spacing of 1.3 nm. Networks larger than 144 nodes do not continue to fit on a 2 cm × 2 cm chip, so they are not shown. The multi-layer networks save the most insertion loss in propagation through low-loss silicon nitride and reduced crossings in traversing through switches, even though the drop loss of a poly-Si switch is larger. Clearly, multi-layer devices are key in either reducing laser power or increasing the scalability of circuit-switched or otherwise spatial networks, in this case enabling an 8 × 8 network.

5.8.2 Matrix-Crossbar

With elimination of the lossy crossings, topologies that were previously not feasible due to a high insertion loss can now be considered. An enabled topology is a full optical crossbar made from matrix-style switches, which can be seen in Fig. 5.53. This topology utilizes fully non-blocking switches which, when coupled with a one-hop electrical setup network such as a flattened butterfly, can be used for both low setup latency and high-bandwidth optical circuits. One drawback is that it utilizes a large number of broadband rings, $O(N^2)$. Therefore, the footprint of the devices must be considered to ensure that all components will fit on a chip. This consideration is captured by the following inequality:

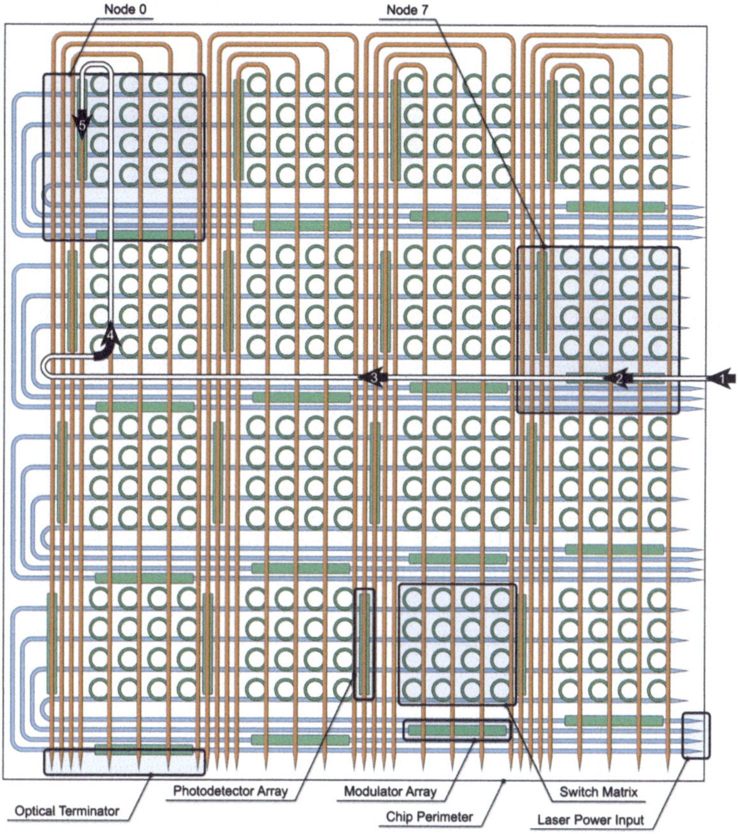

Fig. 5.53 A 4×4 example of the matrix-crossbar network topology (not drawn to scale) [30]

$$S_{chip} \geq N \cdot M \cdot P_{wg} + R_{bend} + N \cdot M \cdot (P_{ring} + D_{ring}) + L_{taper} \tag{5.5}$$

where S_{chip} is the size of the chip dimension (2 cm), P_{wg} is the waveguide pitch (20 μm), R_{bend} is the bending radius for the end loopback-waveguide (20 μm), P_{ring} is the microring resonator pitch (30 μm), D_{ring} is the ring diameter, and L_{taper} is the length of the input taper, or optical terminator. Rearranging the inequality we get a constraint on the maximum size of the microring resonator as:

$$D_{ring} \leq \frac{S_{chip} - R_{bend} - L_{taper}}{N \cdot M} - P_{waveguide} - P_{ring} \tag{5.6}$$

For example, for an 8×8 network, a maximum microring resonator diameter of 261 μm can be obtained. Large microring resonator sizes can be achieved within a small area by using Archimedean spirals to reduce the ring footprint [31].

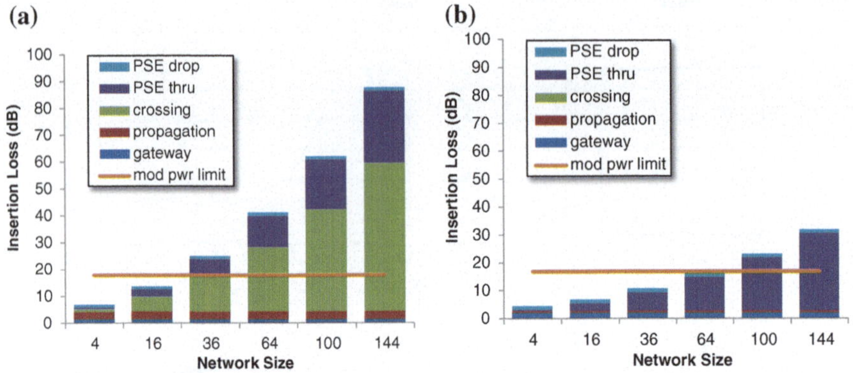

Fig. 5.54 Circuit-switched matrix crossbar insertion loss analysis versus network size [30]

The same insertion loss analysis is performed on this new topology using a channel spacing of 1.8 nm, resulting in Fig. 5.54. This comparison is somewhat contrived since such a single-layer implementation would never be considered due to the copious amounts of crossings that would result. This example illustrates that a totally non-blocking and very high-performance network topology is enabled by the use of multi-layer deposited devices.

References

1. J. Chan and K. Bergman, "Photonic interconnection network architectures using wavelength-selective spatial routing for chip-scale communications," *Optical Communications and Networking, IEEE/OSA Journal of*, vol. 4, no. 3, pp. 189–201, Mar. 2012.
2. G. Hendry, E. Robinson, V. Gleyzer, J. Chan, L. P. Carloni, N. Bliss, and K. Bergman, "Circuit-switched memory access in photonic interconnection networks for high-performance embedded computing," in *International Conference for High Performance Computing, Networking, Storage, and Analysis (Supercomputing)*, Nov. 2010.
3. A. Shacham, K. Bergman, and L. Carloni, "On the design of a photonic network-on-chip," in *First International Symposium on Networks-on-Chip*, 2007.
4. H. Wang, B. G. Lee, A. Shacham, and K. Bergman., "On the design of a 4 × 4 nonblocking nanophotonic switch for photonic networks on chip," in *Proceedings of Frontiers in Nanophotonics and Plasmonics*, 2007.
5. J. Chan, A. Biberman, B. G. Lee, and K. Bergman, "Insertion loss analysis in a photonic interconnection network for on-chip and off-chip communications," in *IEEE Lasers and Electro-Optics Society (LEOS)*, Nov. 2008.
6. A. Shacham, K. Bergman, and L. Carloni, "Photonic networks-on-chip for future generations of chip multiprocessors," *Computers, IEEE Transactions on*, vol. 57, no. 9, pp. 1246–1260, Sep 2008.
7. S. Vangal, J. Howard, G. Ruhl, S. Dighe, H. Wilson, J. Tschanz, D. Finan, A. Singh, T. Jacob, S. Jain, V. Erraguntla, C. Roberts, Y. Hoskote, N. Borkar, and S. Borkar, "An 80-tile sub-100-W teraflops processor in 65-nm CMOS," *Solid-State Circuits, IEEE Journal of*, vol. 43, no. 1, pp. 29–41, Jan.

8. F. Xia, L. Sekaric, and Y. Vlasov, "Ultracompact optical buffers on a silicon chip," *Nature Photonics*, vol. 1, pp. 65–71, Jan. 2007.
9. W. Bogaerts, P. Dumon, D. V. Thourhout, and R. Baets, "Low-loss, low-cross-talk crossings for silicon-on-insulator nanophotonic waveguides," *Opt. Lett.*, vol. 32, no. 19, pp. 2801–2803, 2007.
10. B. G. Lee, A. Biberman, P. Dong, M. Lipson, and K. Bergman, "All-optical comb switch for multiwavelength message routing in silicon photonic networks," *IEEE Photonics Technology Letters*, vol. 20, no. 10, pp. 767–769, May 2008.
11. J. Chan, G. Hendry, A. Biberman, and K. Bergman, "Architectural exploration of chip-scale photonic interconnection network designs using physical-layer analysis," *Lightwave Technology, Journal of*, vol. 28, no. 9, pp. 1305–1315, May 2010.
12. N. Sherwood-Droz, H. Wang, L. Chen, B. G. Lee, A. Biberman, K. Bergman, and M. Lipson, "Optical 4 × 4 hitless silicon router for optical networks-on-chip (NoC)," *Opt. Express*, vol. 16, no. 20, pp. 15 915–15 922, Sep. 2008.
13. M. Uenuma and T. Motooka, "Temperature-independent silicon waveguide optical filter," *Opt. Lett.*, vol. 34, no. 5, pp. 599–601, Mar. 2009.
14. T. Gensty, W. Elsäßer, and C. Mann, "Intensity noise properties of quantum cascade lasers," *Opt. Express*, vol. 13, no. 6, pp. 2032–2039, 2005.
15. Q. Xu, S. Manipatruni, B. Schmidt, J. Shakya, and M. Lipson, "12.5 Gbit/s carrier-injection-based silicon micro-ring silicon modulators," *OSA Optics Express*, vol. 15, no. 2, pp. 430–436, 2007.
16. K. Preston, S. Manipatruni, A. Gondarenko, C. B. Poitras, and M. Lipson, "Deposited silicon high-speed integrated electro-optic modulator," *Opt. Express*, vol. 17, no. 7, pp. 5118–5124, Mar. 2009.
17. P. Lathi, *Modern Digital and Analog Communication Systems*, 3rd ed. Oxford University Press, 1998.
18. G. Hendry et al., "Analysis of photonic networks for a chip-multiprocessor using scientific applications," in *The 3rd ACM/IEEE International Symposium on Networks-on-Chip*, May 2009.
19. H. Wang et al., "ORION: A power-performance simulator for interconnection networks," in *35th International Symposium on Microarchitecture*, 2002.
20. A. Biberman, J. Chan, and K. Bergman, "On-chip optical interconnection network performance evaluation using power penalty metrics from silicon photonic modulators," in *Interconnect Technology Conference (IITC)*, 2010 International, Jun. 2010, pp. 1–3.
21. P. Kumar, Y. Pan, J. Kim, G. Memik, and A. Choudhary, "Exploring concentration and channel slicing in on-chip network router," in *NOCS '09: Proceedings of the 2009 3rd ACM/IEEE International Symposium on Networks-on-Chip*, 2009, pp. 276–285.
22. Cactus Computational Toolkit, [Online]:http://www.cactuscode.org/.
23. Z. Lin, S. Ethier, T. S. Hahm, and W. M. Tang, "Size scaling of turbulent transport in magnetically confined plasmas," *Physical Review Letters*, vol. 88, no. 19, p. 195004, Apr. 2002.
24. A. Canning, L. Wang, A. Williamson, and A. Zunger, "Parallel empirical pseudopotential electronic structure calculations for million atom systems," *Journal of Computational Physics*, vol. 160, no. 1, pp. 29–41, 2000.
25. J. Borrill, J. Carter, L. Oliker, and D. Skinner, "Integrated performance monitoring of a cosmology application on leading HEC platforms," in *Proceedings of the 2005 International Conference on Parallel Processing*, 2005, pp. 119–128.
26. N. Travinin, H. Hoffmann, R. Bond, H. Chan, J. Kepner, and E. Wong, "pMapper: Automatic mapping of parallel Matlab programs," in *DOD_UGC '05: Proceedings of the 2005 Users Group Conference on 2005 Users Group Conference*, 2005, p. 254.
27. H. Kim, E. Rutledge, S. Sacco, S. Mohindra, M. Marzilli, J. Kepner, R. Haney, J. Daly, and N. Bliss, "Pvtol: Providing productivity, performance and portability to DoD signal processing applications on multicore processors," in *HPCMP-UGC '08: Proceedings of the 2008 DoD HPCMP Users Group Conference*, 2008, pp. 327–333.

28. V. Strassen, "Gaussian elimination is not optimal." *Numerische Mathematik*, vol. 14, no. 3, pp. 354–356, 1969.
29. J. W. Cooley and J. W. Tukey, "An algorithm for the machine calculation of complex fourier series," *Mathematics of Computation*, vol. 19, pp. 297–301, 1965.
30. A. Biberman, K. Preston, G. Hendry, J. Chan, N. Sherwood-Droz, J. Levy, M. Lipson, K. Bergman, "Photonic network-on-chip architecture using multi-layer deposited silicon materials for high-performance chip multiprocessors." *Journal of Emerging Technologies in, Computing Systems*, 2011.
31. D.-X. Xu, A. Delâge, R. McKinnon, M. Vachon, R. Ma, J. Lapointe, A. Densmore, P. Cheben, S. Janz, and J. H. Schmid, "Archimedean spiral cavity ring resonators in silicon as ultra-compact optical comb filters," *Opt. Express*, vol. 18, no. 3, pp. 1937–1945, 2010.

Chapter 6
Photonic Network Architectures II: Wavelength Arbitration and Routing

Wavelength-routed networks use individual wavelengths which can be statically or dynamically allocated to source-destination pairs using combinations of modulators, filters, and waveguides. Wavelength-routed networks use wavelength selectivity in order to route data through the network, in contrast to circuit-switched networks which utilize wavelength selectivity for bandwidth aggregation. The wavelength arbitration technique limits the point-to-point bandwidth to a subset of the total number of wavelengths available in the system. This chapter considers architectures where each point-to-point link is composed of a single wavelength channel (unless specified) for simplicity of discussion. This assumption is true of most wavelength-routed networks that have been proposed in literature. These architectures typically exhibit lower latencies than circuit-switched architectures since they do not require the path-setup protocol. Fundamentally, the latency of wavelength-routed networks is only limited by the speed of light and the time required to perform CDR. This chapter describes some of the fundamental building blocks of wavelength-routed architectures and explores some examples of architectures that have been proposed.

6.1 Wavelength Bus Structures

The wavelength bus is the main structure of data transportation, usually consisting of waveguides that intersect with all access points in a network. To accomplish this in a tiled array of access points which is typical of a chip composed of a grid of processing cores, a serpentine layout can be employed to avoid waveguide crossings. Various building blocks are typically used in wavelength-arbitrated designs, including destination-routed buses, source-routed buses, Multi-Write Single-Read (MWSR) buses, Single-Write Multi-Read (SRMW) buses, and a token arbitration ring. These structures are described in the following subsections.

K. Bergman et al., *Photonic Network-on-Chip Design*,

Integrated Circuits and Systems, DOI: 10.1007/978-1-4419-9335-9_6,

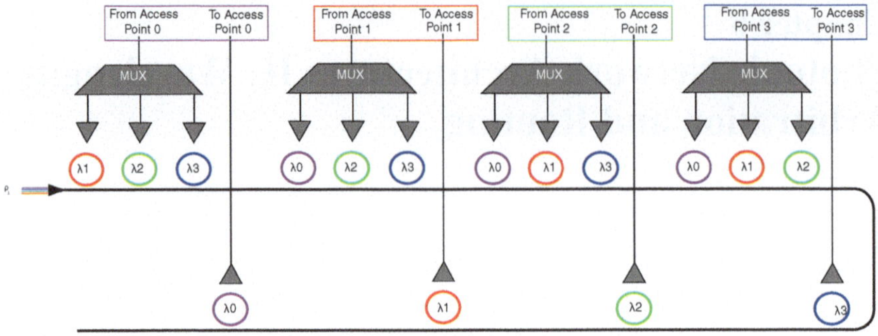

Fig. 6.1 A source-routed wavelength bus connected to four access points. Each access point receives data on a dedicated wavelength channel and transmits on all other wavelength channels

6.1.1 Source-Routed Bus

The source-routed bus is a configuration where each access point reads from a single wavelength channel that has been assigned to it. Any other access point can write to this channel. Figure 6.1 shows a possible implementation with microring modulators, ring filters, and detectors. The layout uses N^2 rings and N wavelength channels, where N is the number of access points. Each access point has a single detector that is connected to a ring filter, where the ring filter is tuned to its assigned wavelength channel. Each access point also owns $N-1$ ring modulators which are each tuned to the wavelength channels assigned to all other access points. More bandwidth can be achieved by using higher serial data rates per channel serial data rates or assigning several wavelengths to each access point. Scaling the number of wavelength channels per access point is limited by the scale of the network, the channel spacing, and the modulator FSR.

In order to prevent network collision, multiple access points must be disallowed from concurrently writing to a common destination. This could be accomplished with a separate arbitration network, such as a token arbitration ring which is discussed in Sect. 6.1.6. Once arbitrated, a single access point will transmit data to the destination while all other access points must configure their corresponding modulators to an off resonance state to allow modulated signals to pass by undisturbed.

For example, referring to Fig. 6.1, consider the case when access point 0 would like to transmit data to access point 3. First, arbitration is performed to allocate the blue wavelength channel to access point 0 which will be destined for access point 3. At the same time, access point 1 and access point 2 make sure that their modulators are tuned for the blue wavelength are off resonance. The consequence of not driving the extra modulators off resonance is that they would interfere with the signal before being received by access point 3.

The disabling of non-communicating modulators can be accomplished by designing each ring to be off resonance when it is not being driven by a voltage. This effectively places the microring modulator in an off resonance state by default.

This is advantageous since it avoids having to actively communicate to every access point every transaction that occurs on the network. This disabled state is not necessarily the same as when the microring modulator is placed in an off-resonance state for transmitting a '1' bit. The off-resonance state for '1' bit transmission and the on-resonance state for '0' bit transmission will drive the ring with a particular voltage that places the resonant modes at predefined wavelengths. These two data states must be far enough apart to ensure adequate extinction ratio while at the same time being close enough together to allow a specified modulation data rate. A balance must be met since a short resonance shift will result in a high possible data rate but low extinction ratio (which will result in bit errors), while a large resonance shift will have a larger extinction ratio (fewer bit errors) but will be harder to drive at high data speeds.

The default off resonance behavior in microring modulators for enabling a disabled state raises several technical challenges. First, if all microring modulators are off resonance, then the optical power will reach the detectors when there is no activity. Coding or other data-valid signaling must be implemented to properly detect data, and the receiver circuits must be designed for low leakage power when amplifying a logical '1' to avoid burning unnecessary power when the network is inactive. A second concern is that the resonant modes of the modulator in the default disabled state must be spaced far enough from the wavelength channel to minimize the insertion loss experienced by optical messages that are passing by. This disabled state can potentially be farther off resonance than the '1' bit off resonance state. This potentially creates three states the modulator must be able to achieve, which would require more complex driving circuitry.

6.1.2 Destination-Routed Bus

The dual of the source-routed wavelength bus is the destination-routed wavelength bus. Figure 6.2 shows the implementation using ring resonator microring modulators, tunable ring filters, and photo-detectors. In this configuration, an access point modulates a single wavelength channel, and its intended destination selectively reads it. Like the source-routed implementation, this method uses N^2 rings and N wavelength channels.

In order to determine the intended destination, arbitration must occur *a priori* using the token arbitration ring discussed later in Sect. 6.1.6. The filters in this design are similarly manipulatable like the electro-optic switches discussed in Sect. 3.5, however they only operate on a single wavelength channel. To explicitly make this distinction, we refer to the electro-optic filters as *tunable filters*. The tunable filters of each access point are tuned to a default off resonance state. When an optical message must reach a particular destination, the appropriate tunable ring filter is adjusted to an *on*-resonance state. Unlike the source-routed wavelength bus technique, the tunable filters used here only need to have the ability to be in two states, reducing the complexity of the driver circuitry.

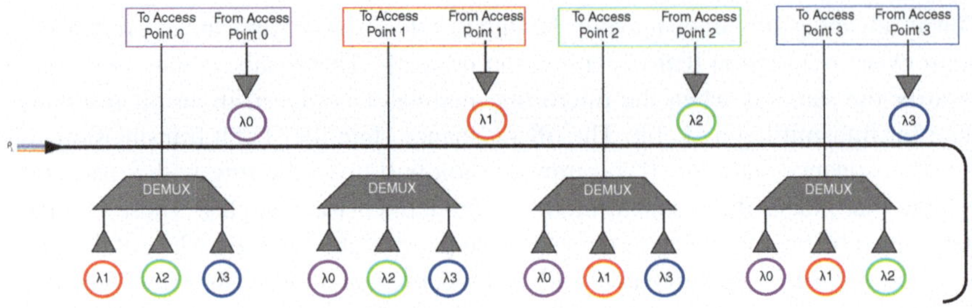

Fig. 6.2 A destination-routed wavelength bus connected to four access points. Each access point transmits on a dedicated wavelength channel and receives on all other wavelength channels

An alternative method for arbitration of the destination-routed bus is to use the broadcast and select technique. This is based on non-tunable filters, which are designed to tap off a fractional amount of the power so that all access points receive the same transmission in a broadcast fashion. An encoded destination identifier tag is read from the message header and compared to the identifier of the destination access point. The message is discarded if the identifiers do not match. This requires more power per wavelength channel than the other methods, since a portion of tapped power must be enough to satisfy the detector sensitivity. This method also requires very high precision and stability of the ring filter resonant modes since they each must tap off a small portion of the total optical power. If present, small drifts in resonance of any of the filters will highly disrupt the operation of the whole bus.

6.1.3 Multi-Write Single-Read

A third way of connecting access points together is using a multi-write single-read (MWSR) bus. Though not strictly wavelength-routed, networks designed to use this structure often have a similar serpentine or other layout which passes by every access point. The configuration can be seen in Fig. 6.3. Each destination has its own dedicated waveguide, with a set of $N - 1$ filters tuned to all wavelength channels. Each access point can transmit on all wavelengths on the waveguides on all other access points. Note that Fig. 6.3 only shows the waveguides for access point 3 (top bus) and access point 2 (bottom bus). Access point 0 and access point 1 also have separate waveguides (not shown), with every other access point possessing a set of modulators to send to it. Clearly, MWSR must also be arbitrated *a priori*, and must have default off resonance modulators. In the MWSR scheme, however, bandwidth is more easily achieved through WDM because the wavelength domain is not used for routing, such as in Fig. 6.3 where 3 wavelengths are aggregated together to increase total throughput. The MWSR scheme uses M wavelengths per waveguide and N waveguides, where N is the number of access point. Each point is connected to $M \times (N - 1)$

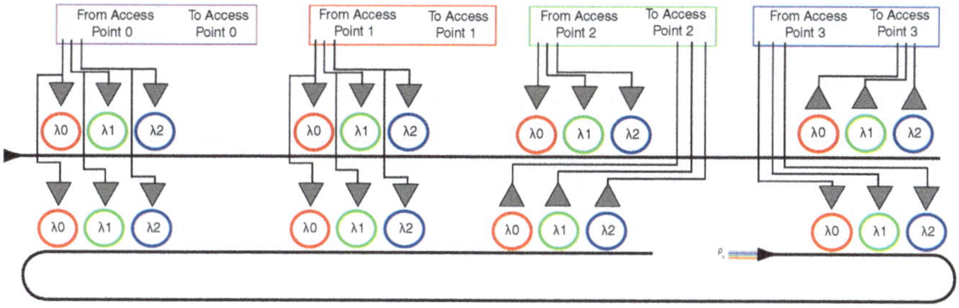

Fig. 6.3 A Multi-Write Single-Read wavelength bus connected to four access points. Each access point is assigned a waveguide and is capable of receiving on all (three) wavelength channels on that waveguide. Each access point is also connected to every other waveguide and capable of transmitting on all wavelength channels on those waveguides. Only two of the total four waveguides are shown in this figure

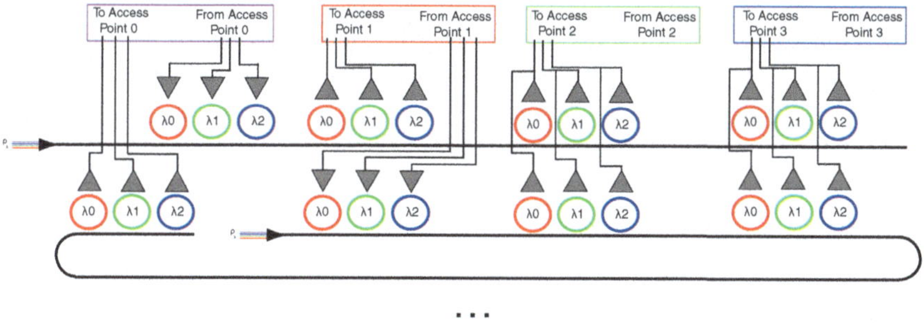

Fig. 6.4 A Single-Write Multi-Read wavelength bus connected to four access points. Each access point is assigned a dedicated waveguide and capable transmitting on all wavelength channels on that waveguide. Each access point is also capable of receiving wavelengths on all other waveguides in the network

modulators and M photo-detectors. Therefore, each waveguide possesses $M \times N$ rings while the network uses a total of MN^2 rings.

6.1.4 Single-Write Multi-Read

The dual of MWSR is the single-write multi-read (SWMR) wavelength bus, shown in Fig. 6.4. Like MWSR, SWMR has a separate waveguide per access point which is routed past all other access points. In SWMR, each access point possesses a single set of modulators on its own dedicated waveguide. Each access point is also connected to a set of tunable ring filters and photo-detectors on all other waveguides. The tunable filters are shifted on resonance once arbitration is complete. In SWMR, each access point is connected to M modulators and $M \times (N - 1)$ photo-detectors. $M \times N$ rings are placed on each waveguide and the network uses a total of MN^2 rings. The

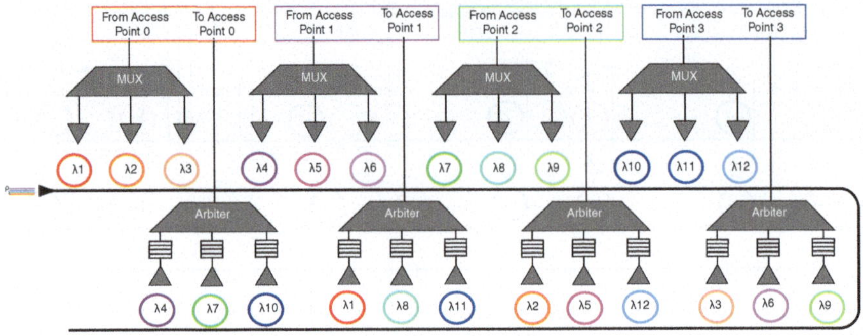

Fig. 6.5 A wavelength crossbar connecting to four access points. Each source-destination combination uses a dedicated wavelength for communications

SWMR bus offers one advantage with respect to the MWSR bus: if the filters can be designed such that they only tap off a small amount of the optical power, then broadcast-and-select functionality can be used.

6.1.5 Wavelength Crossbar

A disadvantage of the wavelength arbitration schemes is that they either require *a priori* arbitration or a complex broadcast-and-select mechanism. To avoid having to arbitrate, we can combine the source-routed wavelength bus and destination-routed wavelength bus techniques into a full wavelength crossbar, as shown in Fig. 6.5. Here, each source modulates data on a different wavelength channel depending on the destination, and each destination receives on a different wavelength depending on the source. Therefore each source-destination pair utilizes a dedicated wavelength channel. Since an access point can now receive concurrent transmissions from multiple sources, messages must be appropriately organized and appropriately managed before forwarding it to the compute node. This implementation requires $2N \times (N-1)$ rings and $N \times (N-1)$ wavelength channels.

Unlike wavelength arbitration techniques, the wavelength crossbar does not require any tuning or detuning of microring modulators or ring filters. This would simplify the driver circuit for the modulators and/or filters. A disadvantage of this design is that the wavelength crossbar does require significantly more wavelengths: $N(N-1)$ for an N-node network. For any reasonable sized network, it is not feasible to put this many wavelengths on the same waveguide. A workaround is to partition the set of source-destination pairs and use multiple waveguides.

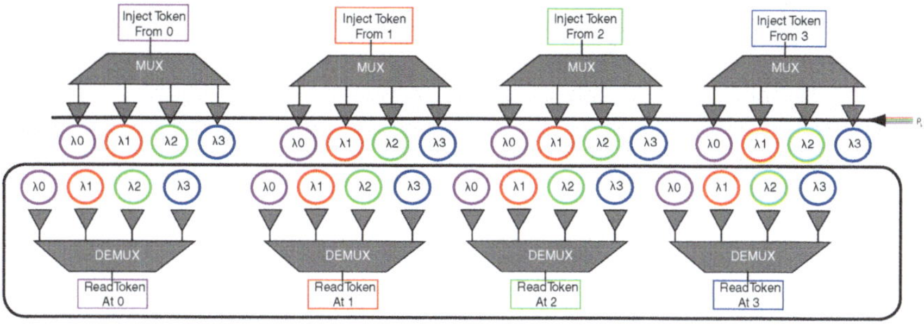

Fig. 6.6 Token arbitration ring used for optical arbitration. This token ring is able to arbitrate four different resources (one per wavelength channel) for four different network access points

6.1.6 Token Arbitration Ring

The optical token arbitration ring implements a method for allocating wavelength channels in a wavelength-routed network to prevent data collision [1]. This structure, illustrated in Fig. 6.6, allows access points to pass around tokens in order to gain control access to particular resources. Only an access point that has ownership of a token can utilize the resource assigned to the token. For instance, permission to transmit to access point 2 on λ_2 in the source-routed wavelength bus in Fig. 6.1 can be implemented with a token. An access point must relinquish a token by placing it back on the token ring so that another access point can utilize the resource. Each network resource that requires provisioning is assigned a token, and each token is assigned a corresponding wavelength channel in the token arbitration ring. The token is taken and placed on the network in such a way that only a single access point can utilize the associated resource at a time.

The arbitration ring is composed of a looped waveguide and a power waveguide. Each access point has a set of modulators and a set of ring filters with photo-detectors, equal to the number of tokens used in the network. Each microring modulator is tuned to one of the token wavelengths. Similarly, the ring filters that are connected to the photo-detectors are each tuned to a token ring wavelength. The power waveguide serves as a CW light delivery mechanism to each tunable ring filter. If an access point wants to inject a token into the token ring, the tunable filter is adjusted to couple light from the power waveguide onto the token ring.

Each optical token is designated a corresponding resource on a network (e.g. a wavelength channel). Consider a situation where access point 0 injects λ_0 (purple) from the power waveguide. If access point 2 wants to send data to access point 0, it shifts its λ_0 ring filter into the on-resonance state to capture the optical signal. If access point 2 is receiving λ_0, that means that no downstream access point can receive the token because the read is destructive. Therefore, access point 2 now has exclusive access to access point 0. Once access point 2 finishes using the resource, it can release the token by re-injecting it from the power waveguide. By default, each access point has the filter for its own wavelength channel set to the on-resonance state, so that it can re-send its own token if no access point claims it. A token ring that uses N tokens requires $2N^2$ rings and N wavelengths, and requires tunable filters for both token injection and reclaiming tokens.

Reference

1. D. Vantrease, R. Schreiber, M. Monchiero, M. McLaren, N. P. Jouppi, M. Fiorentino, A. Davis, N. Binkert, R. G. Beausoleil, and J. H. Ahn, "Corona: System implications of emerging nanophotonic technology", in Proceedings of 35th International Symposium on Computer, Architecture, Aug. 2008.

Chapter 7
Photonic Network Architectures III: Advanced Photonic Architectures

Circuit-switching is a natural use of the photonic transmission medium because it is a simple solution that abstracts many of the physical layer implementation details in network design. However, it can suffer from severe network congestion and starvation when long-lived communication patterns occur. This chapter explores some architectural variations on circuit-switched networks to alleviate some of these issues.

7.1 Time Division Multiplexed Arbitration

Although optical circuit-switched links are very efficient at transporting data across a chip when they are established, the path-setup protocol does not have any notion of fairness. This lack of fairness can lead to degraded performance under high loads due to contention. One option to address this problem is to implement a time-division multiplexing (TDM) arbitration and distributed control of photonic switches [1]. The basic concept is the follows: during a specified amount of time, or time slot, switches in the network are configured to allow communication between one or more pairs of access points. Each time slot is of length

$$t_{slot} = t_{setup} + t_{transmission} + t_{propagation} \tag{7.1}$$

where t_{setup} is the time it takes to change the state of a ring resonator, $t_{transmission}$ is the time each node is allowed to transmit data per time slot, and $t_{propagation}$ is the worst-case propagation latency between any two valid communicating pairs. If each switch can keep track of the current time slot using a global clock, it can be made aware of its correct configuration using control registers for any given time slot. This allows the control of the switches to be completely distributed.

This concept should be distinguished from traditional TDM mechanisms in other networks. Typically, requests to use network resources are arbitrated by sources or

K. Bergman et al., *Photonic Network-on-Chip Design*,
Integrated Circuits and Systems, DOI: 10.1007/978-1-4419-9335-9_7,

individual network nodes to dynamically allocate a temporal schedule for access to virtual channels, physical links, switches, or virtual circuits, thus providing fairness guarantees to latency and bandwidth [2–6]. The method shown in this section aims at providing the same fairness, but because there is no practical equivalent to a buffer implementation in silicon photonic integrated technology, TDM arbitration must be applied through the entire network by creating end-to-end optical circuit paths. Here, the scheduling of node accesses to network resources can be done statically at design time, or potentially at chip-power-on or even application startup. If there are N_{slot} time slots, each of duration t_{slot}, then the total time for a complete TDM cycle is

$$T_{TDM} = N_{slot} \times t_{slot} \tag{7.2}$$

This complete set of TDM slots is called a TDM frame, and T_{TDM} is the TDM frame time. Two variants of the TDM concept are applied for photonic networks in the following sections. The first variant is the fully-connected TDM (Sect. 7.1.1) which considers a photonic TDM network that provides a completely optical end-to-end transmission. The second variant is the Enhanced TDM (Sect. 7.1.2) which introduces an intermediate stage of optical-electronic-optical conversion in the source-destination data transmission.

7.1.1 Fully-Connected TDM Arbitration

Fully-connected TDM arbitration stipulates that every network node must have the opportunity to send messages to every other node within T_{TDM}.

In this arbitration scheme, the network repeatedly cycles through every time slot. If a network node has data to send to another node, it waits for the correct time slot. If a node has multiple messages to different destinations queued up, it can send them out of order. Also, by statically selecting different values for $t_{transmission}$, the granularity of the arbitration can be varied. If, for instance, the system architecture specifies that only fixed-length messages may be sent on the network (i.e. cache lines), then $t_{transmission}$ can be adjusted to exactly match that size.

The naive way to accomplish the resource scheduling is to assign a single time slot to every communicating pair in the network. Thus, the network would require

$$N_{slot} = N \times (N - 1) \tag{7.3}$$

time slots to implement full coverage, where N is the number of nodes in the network. A 64-node network would therefore require 4032 time slots. This naive scheduling of one transmission per time slot in the network achieves the worst-case network utilization.

The naive implementation can be improved by scheduling more than one transmission per time slot, thus reducing the total number of time slots, and the worst-case latency of a message waiting for its slot. In order to maintain correct operation, the implementation must adhere to the following constraints during a single time slot:

1. **Source contention.** A node can only send to one destination, assuming a single set of modulators at an access point.
2. **Destination contention.** A node can only receive from one source, assuming a single set of detectors at an access point.
3. **Topology contention.** Transmission cannot overlap in the same waveguide.

These constraints present the problem of scheduling in both time and space at least one transmission from every node to every other node in the network. One solution to this scheduling problem is to search the solution space using a *genetic algorithm* [1]. A 4×4 mesh, for example, requires only 18 time slots. The combination of TDM with concentration, which is discussed in Sect. 5.4.1, enables the design of smaller networks with lower power dissipation (smaller optical losses lead to smaller laser power dissipation) as well as better performance (thanks to the smaller number of time slots).

7.1.1.1 TDM Switch

TDM-arbitrated networks are characterized by some implementation changes with respect to pure circuit-switching networks. Figure 7.1 shows the layout of a photonic switch in the TDM network. It consists of waveguide paths and assumes electro-optically controlled 200-μm ring resonator-based PSEs, which spatially switch a broadband signal as a purely circuit-switched network would. Ports are labeled as North, South, East, West, and Gateway.

This implementation is optimized for X-then-Y dimension-order routing, therefore the switch does not need to implement full connectivity between the ports. Table 7.1 shows the port combinations, and the PSE number that implements the path, referring to Fig. 7.1. For example, the PSE labeled as 1 can switch a signal from the gateway (modulator bank) to the north port. Note that the signal must pass through a ring only when coming from a gateway, entering a gateway, and turning from an X-dimension to Y-dimension. Hence, there is a savings on insertion loss when signals travel in straight lines.

7.1.1.2 Switch Controller

In the network architecture, each switch is controlled by a local controller which is aware of the current TDM slot by tracking ticks of a global TDM clock and is therefore aware of how the switch should be set. A global, synchronous TDM

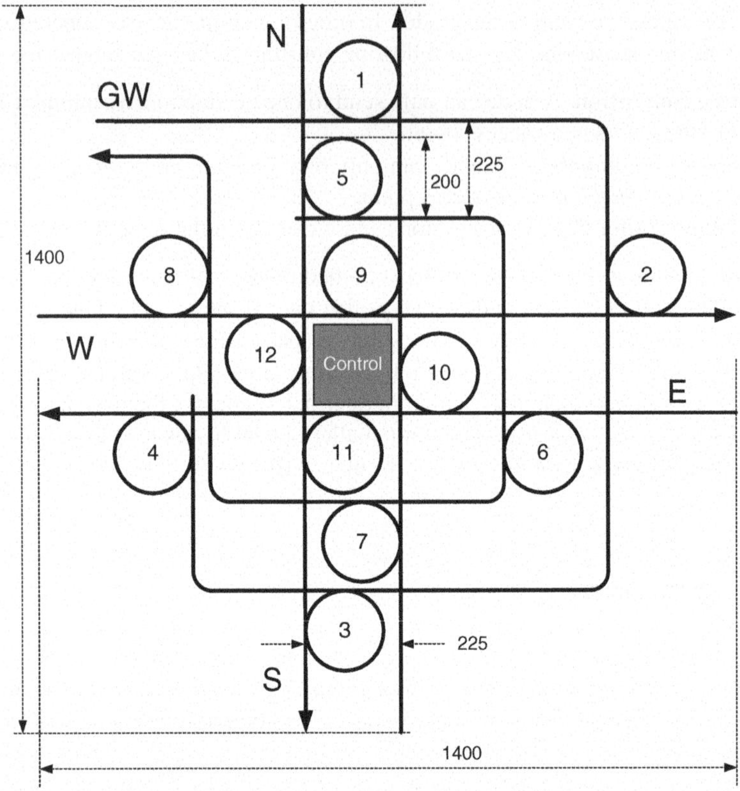

Fig. 7.1 Layout of TDM photonic switch, showing waveguides and ring resonators. Units in microns [7]

Table 7.1 Switch functionality [7]

Inport	Outport	PSE
Mod	N	1
Mod	E	2
Mod	S	3
Mod	W	4
N	Det	5
E	Det	6
S	Det	7
W	Det	8
W	N	9
E	N	10
E	S	11
W	S	12
E/W	W/E	–
N/S	S/N	–

clock can be implemented with waterfall clock distribution, synchronous latency-insensitive design [8], or optical clock distribution [9]. The period of this clock must be the TDM period, t_{slot}. Clearly, t_{slot} should be set to an expected average message transmission time, so that time slots are just big enough to allow end-to-end transmission within a single TDM slot. Taking into account time of flight overhead, this value could be tens of nanoseconds equating to less than 250 MHz TDM clock frequency (depending on t_{slot}), a very feasible implementation by today's standards.

The output logic can be implemented as a single lookup table (LUT) which takes the switch ID register as an input. In practice, only the fraction of the table that is necessary to run the local switch would be instantiated to save area and power. The size of the output logic is proportional to the number of TDM slots, which is dictated by the number of network nodes. Specifically, there is one bit per PSE per TDM slot, indicating whether the PSE is on or off. Since there are 12 PSEs per switch, the LUT memory must contain $1.5 \times N_{slot}$ bytes of information for each switch controller.

7.1.1.3 TDM Network Gateway

Figure 7.2 shows the microarchitecture of a network gateway, providing network and memory access to four cores. This is done by a main *TDM controller*, which

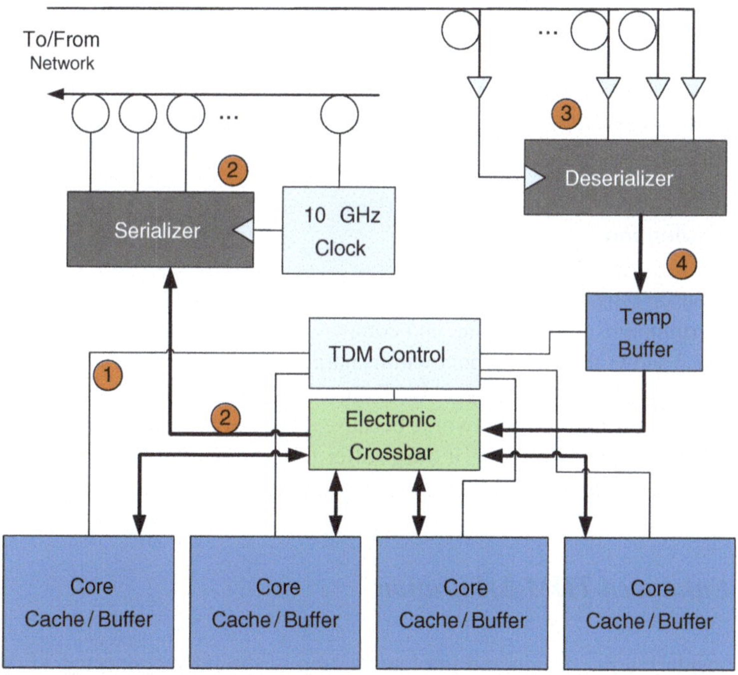

Fig. 7.2 TDM network gateway microarchitecture [7]

arbitrates access to the network. The gateway operation consists of five main steps, numbered in Fig. 7.2:

1. Communication requests are made to the TDM controller, which controls an electronic crossbar that connects the various gateway components.
2. When the network is in the correct TDM slot, the TDM controller sets the crossbar from the requesting core to the serializer, which ramps the data up to the modulation data-rate (shown as 10 Gb/s). This bitrate clock can also be transmitted on a separate wavelength for data recovery at the receivers.
3. When a signal is received, it is deserialized, clocked by the received transmission clock.
4. If the data has reached its destination, it sits in a temporary buffer, waiting for access to the electronic crossbar. Access will be immediately available unless cores in the same gateway are communicating locally through the crossbar.

The temporary buffer is only used to store received transmissions that are destined for the cores in the gateway. The TDM controller gives priority to the temporary buffer over local core-core communication. Therefore, it needs to hold a maximum of 3 transmissions: one for receiving incoming transmissions, one for sending the last received transmission on to the correct core, and one buffer in case of destination contention.

One important function of the gateway in the TDM architecture is to allow *out-of-order* access to the network from the cores, a property inherent in the TDM architecture. In other words, if a request from one core can be sent during a time slot, it does not have to wait for other requests from other cores that need other time slots to be serviced even if their requests arrive to the controller first. This property motivates the microarchitecture design decision of using concentration to increase the network utilization.

7.1.1.4 Evaluation

This section investigates the case study of a 4×4 TDM network with 4-way concentration running Random traffic, and compare it to an E-Mesh and P-Mesh.

Figure 7.3 shows the latency/bandwidth characteristics for small (256 B), medium (8 kB) and large (256 kB) messages. In all cases, the TDM network achieves higher bandwidth than the other two networks because it accomplishes its goal of providing some level of fairness. However, it does suffer from higher zero-load latencies due to messages having to wait for their time slot to begin transmission.

7.1.2 Enhanced TDM Arbitration

After considering the design and implementation of the fully-connected TDM network, it can be seen that it is possible to decrease the number of time slots required, as described by Hendry et al. [10].

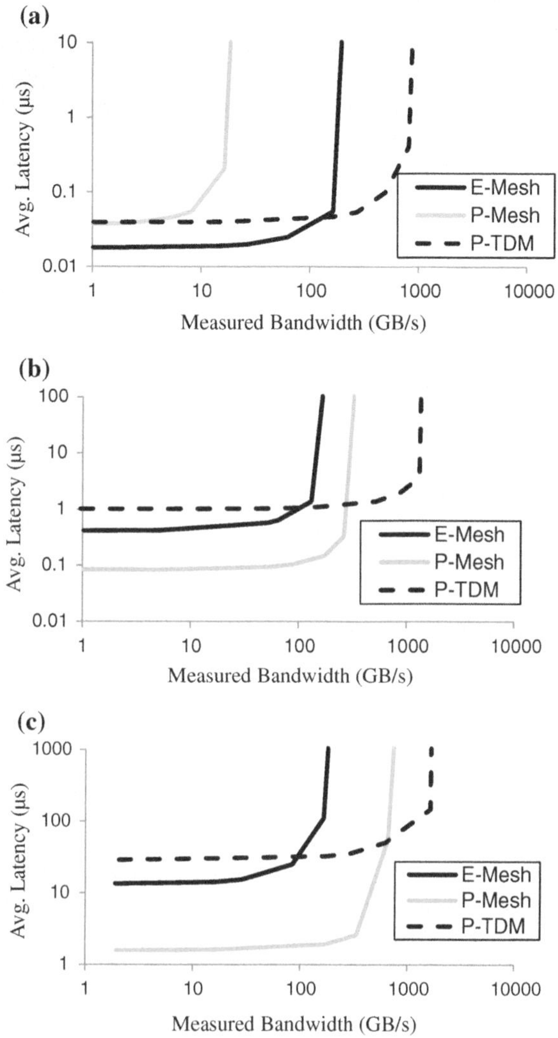

Fig. 7.3 Latency and bandwidth characteristics for a full-coverage 4 × 4 TDM network concentrated to 64 access points for (**a**) 256 B messages, (**b**) 8 kB messages, (**c**) 256 kB messages [7]

To simplify the problem, the arbitration scheme is modified so that photonic transmission will no longer be entirely end-to-end for every node pair. Rather, the mesh X-dimension transmission is first completed, converted to the electronic domain, and stored in a buffer until the Y-dimension transmission can be completed. This means that optical to electrical conversion energy is expended twice, once at the intermediate X-Y turning point and a second time at the destination. This simplification reduces the energy per bit benefits that end-to-end photonic transmission technology provides. This segmentation of X and Y direction communications is referred to as enhanced TDM arbitration (ETDM).

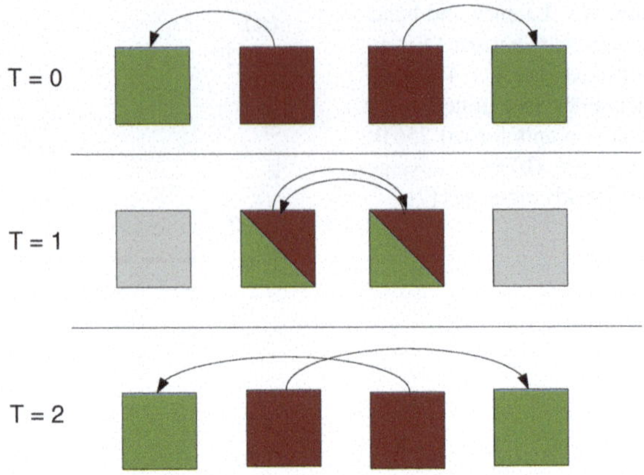

Fig. 7.4 Row communication TDM slots example for four nodes [10]

Two transmissions can always take place in a row during a time slot, for any size row, where the two sending nodes are on opposite sides of the row. This is illustrated in Fig. 7.4 for one row of four nodes, assuming bidirectional links connecting neighboring nodes consisting of two uni-directional waveguides. The red nodes are the sending nodes, and exhaust all possible combinations of destinations (green) in the row. The process repeats for all other nodes being designated as the sending nodes. Note that communications are shown symmetric across the midpoint of the row in Fig. 7.4, though this is not required.

Communications can then be scheduled such that two transmissions occur in every row and every column in each time slot. Since each node in a row must communicate with every other in its row, and two nodes are communicating at once per row, this requires

$$N_{slot} = (R - 1) \times \binom{R}{2} \tag{7.4}$$

time slots, where R is the number of nodes in a row (and column, assuming a square network), R is even, and $R \geq 4$. For an 8×8 64-node network, this is merely 28 time slots, a significant improvement over the previous end-to-end implementation with 142 time slots [1].

Figure 7.5 illustrates an example of how to schedule a 4×4 TDM network, which requires 6 time slots. The transmission possibilities are represented as a 16×16 *control matrix*. Each entry in the matrix is color-coded to indicate which sender-receiver pair is enabled during a time slot. Note that a node may only send and receive once per time slot, which translates into the rule that a color may only appear once in a row and column in the control matrix. Also note that not all node combinations are necessary since the design is simplified to only require optical circuit paths to travel in one mesh dimension during a slot. This simplification is why many entries are blank (white).

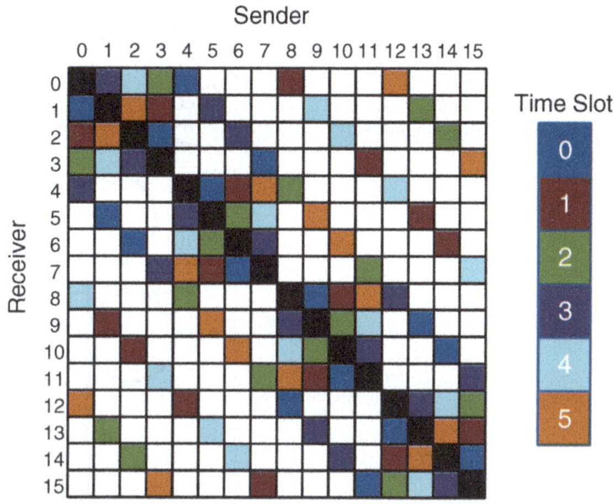

Fig. 7.5 E-TDM control matrix for a 4 × 4 network [10]

Some visual and numerical patterns are useful when specifying the control matrix for any size network. For instance, the 4 × 4 squares lying on the black diagonal indicate row communications. Other diagonal stripes represent column communication. In constructing this matrix, each block (or row in the mesh) utilizes every time slot exactly twice, per Fig. 7.4. The block pattern shifts slightly to accommodate column communications, and is mirrored across the network bisection line (row $R/2$).

7.1.2.1 ETDM Switch

The switches used in enhanced TDM arbitration have several differences from the design of the switch used in fully-connected TDM arbitration. Figure 7.6 shows the layout for the photonic switches in the network.

The TDM period schedule is optimized for fewer TDM slots at the cost of paying additional O-E-O conversion energy. A consequence of the X-then-Y routing is that the switch no longer needs to implement as many paths as the original TDM switch. An optical signal only drops into a ring when entering or exiting a gateway, which saves on insertion loss when traveling straight through the switch.

7.1.2.2 Network Gateway

The ETDM gateway microarchitecture also differs from the design of the fully-connected TDM gateway. Local X-Y buffers are added to the gateway to facilitate the intermediate conversion. Figure 7.7 shows the microarchitecture, providing network and memory access to four cores. This is accomplished through the use of a main *TDM controller*, which arbitrates network and memory resources and acts as a memory controller by keeping a master schedule of events that occur during each time slot.

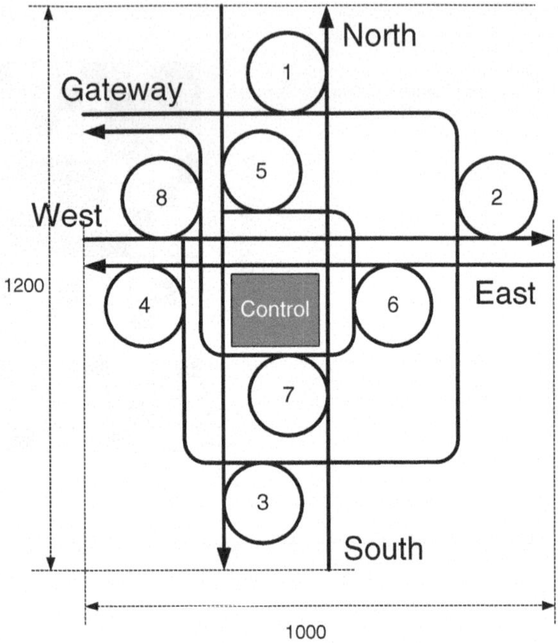

Fig. 7.6 Layout of ETDM photonic switch, showing waveguides and ring resonators. Units in microns [10]

Each gateway has two vertically-coupled connections to a memory bank. Local reads and writes are serviced by scheduling row and column accesses during free slots in the master schedule. Remote memory accesses are sent to the destination gateway, where they are then scheduled in a similar fashion. Remote reads are read directly from memory into the network to save on buffering power.

The following describes an example of the gateway operation, numbered in Fig. 7.7:

1. Communication requests are made to the TDM controller, which controls an electronic crossbar that connects the various gateway components.
2. When the network is in the correct TDM slot, depending on the type of communication (memory read, memory write, network-send, etc.), the TDM controller sets the broadband rings that control access to and from the modulators and detectors. This can also be done ahead of time when the time slot switches, if the transaction has been queued up.
3. The TDM controller also sets the crossbar from the requesting core to the serializer, which ramps the data up to 10 Gb/s modulation. The transmission clock is also transmitted on a separate wavelength channel.
4. When a signal is received, it is deserialized, clocked by the received transmission clock.
5. If the data has reached its destination, it sits in a temporary buffer, waiting for access to the electronic crossbar. Access will be immediately available unless cores in the same gateway are communicating locally through the crossbar.

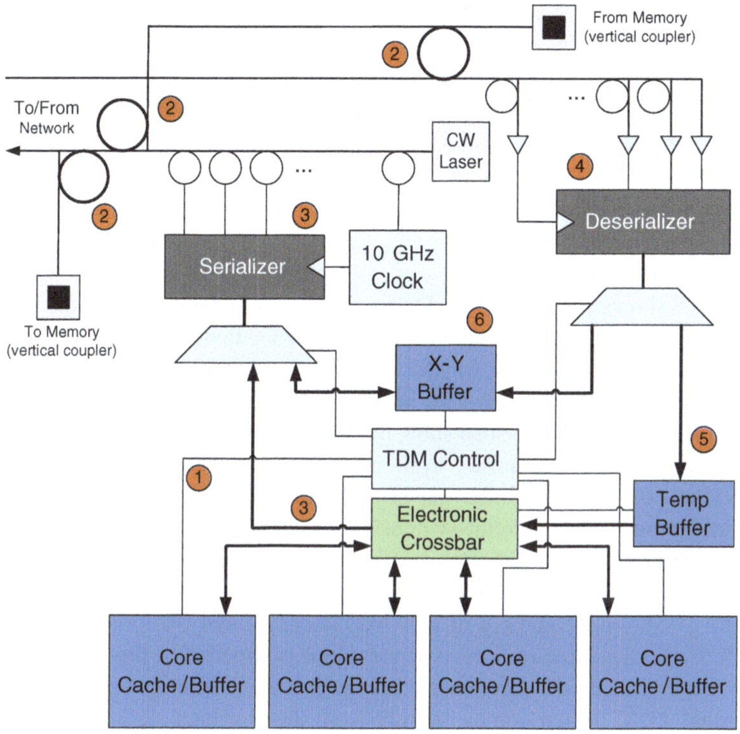

Fig. 7.7 ETDM network gateway microarchitecture [10]

6. If the data is using the gateway as an intermediate point while switching dimensions, it sits in the X-Y buffer and notifies the TDM controller. It can then transmit during the correct TDM slot.

The required size of the buffers can be determined exactly based on the size of the network. The X-Y buffer is used to hold transmissions that have arrived at this gateway to continue through the network in a different dimension, and are waiting for their time slot. Therefore, they must hold a maximum of $2 \times (R - 1)$ transmissions, which is the number of time slots in one TDM frame in which a message could be received. A 64-node network will therefore require a buffer of size $14 \times S_{transmission}$, where $S_{transmission}$ is the maximum message size that can be transmitted in one time slot.

The temporary buffer is only used to store received transmissions that are destined for the cores in the gateway. The TDM controller gives priority to the temporary buffer over local core-core communication. Hence, it needs to hold a maximum of 2 transmissions: one for receiving incoming transmissions, and one for sending the last received transmission on to the correct core.

7.1.2.3 Evaluation

Figure 7.8 reports the results of an experiment to test the ETDM network characteristics using the PhoenixSim design environment presented in Chap. 4. In this experiment the network was simulated for 5 ms by running Random, Tornado, and Hotspot traffic in the network with 8B, 128B, and 2kB messages, representing control, cache-line, and application-level message sizes, respectively. The value of t_{slot} is set at 10 ns, requiring 1 ns each for t_{setup} and $t_{propagation}$, making $S_{transmission}$ equal to 10240 bits, or about 1.2 kB.

Figure 7.8 shows the average read latency versus total bandwidth in the network. The two TDM networks show higher zero load latency than the P-mesh, as expected from the overhead of waiting for the correct slot. However, the enhanced TDM network shows significant zero-load latency improvement over the original TDM design. Both TDM networks also show higher throughput compared to the P-mesh for all message sizes, due mostly to their ability to service message requests that arrive to the gateway's controller out of order, thus increasing network utilization. Bandwidth gains over the circuit-switched solution are most profound in the traffic patterns with more chances of circuit-path blocking in the P-mesh, either from long communication (Random) or predictably conflicting resources (Tornado).

Figure 7.9 shows the sources of zero-load latency under Uniform traffic for each network as message size increases. The P-mesh is superior in this respect, as it is entirely dependent on the electronic router hop latency. The network latency from the original TDM design comes entirely from the slot latency, which is the time spent by a message when it is next in line waiting for a time slot. The ETDM design improves the zero-load latency over the original TDM design by decreasing the time slot count, despite additional delay when changing dimensions (XY-buffer queuing and slot latency). The TDM networks also show a significant increase in latency for the larger 2 kB messages because the message must be sent in multiple slots. Although the slot period could be changed to match the message size for the each simulation to prevent segmentation, a constant slot period is used to illustrate the effects of its relationship to expected message size.

To illustrate the effects of contention on network latency, Fig. 7.10 shows the sources of latency while the P-mesh, P-TDM, and P-ETDM networks are loaded at half capacity (the arrival rate of messages is half the load required to saturate the network). For the P-mesh, blocking latency enters the picture, forcing queuing at the network gateways. The original TDM design is still dominated by slot latency, where queuing latency is dictated by the traffic pattern. The ETDM method has a similar relationship, though much less severe because of the reduced slot count. An extra traffic-dependent queuing latency is introduced at the XY-buffer, though it is small compared to the total.

Figure 7.11 shows a coarse network power breakdown under uniform random traffic near saturation, assuming around 12 % integrated laser efficiency [11]. Electronic circuitry accounts for a large portion of the power dissipated in all three networks. A significant portion of the electronic power dissipation comes from the electronic crossbar necessary to implement external concentration. Because this crossbar must

Fig. 7.8 Latency/bandwidth characteristics of ETDM under synthetic traffic for (**a**) Random, (**b**) Tornado, (**c**) Hotspot [10]

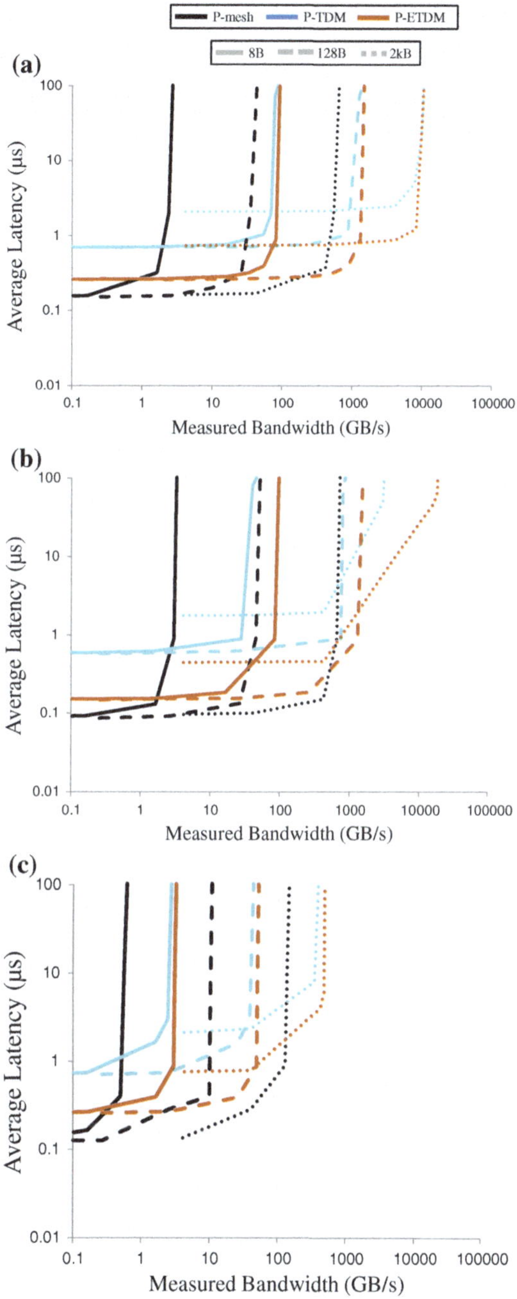

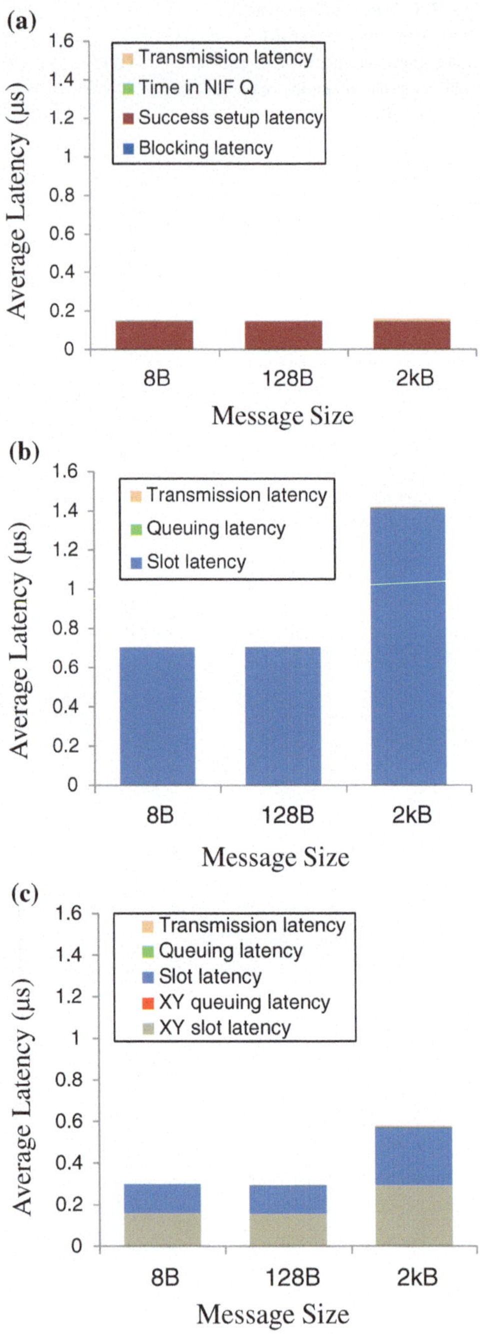

Fig. 7.9 Zero-load latency breakdown under Uniform traffic [10]: **a** P-mesh, **b** P-TDM, **c** P-ETDM

Fig. 7.10 Half-load latency breakdown under Uniform traffic [10]: **a** P-mesh, **b** P-TDM, **c** P-ETDM

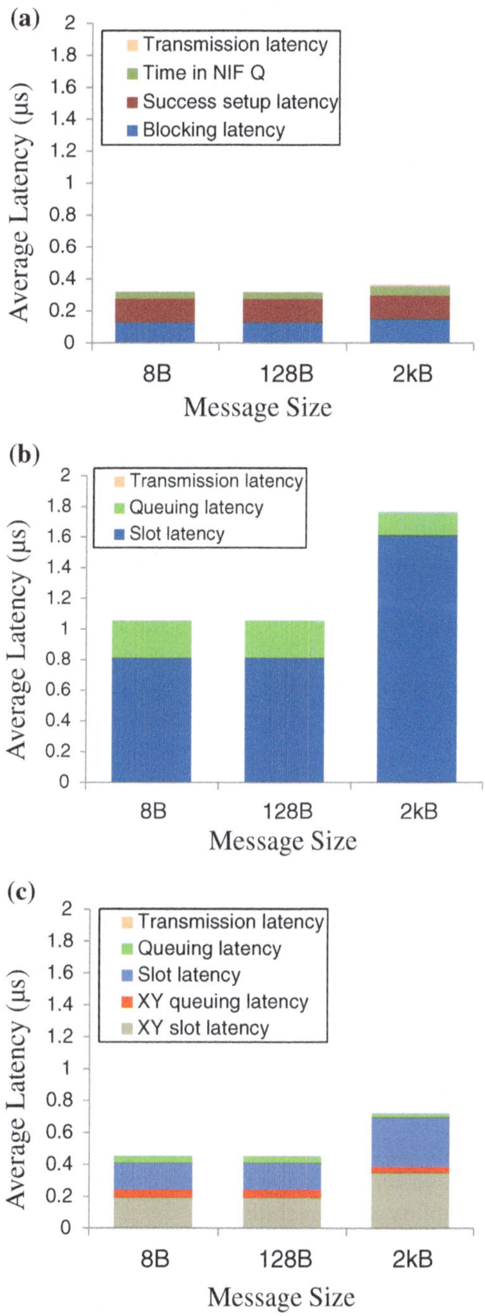

Fig. 7.11 Power breakdown [10]

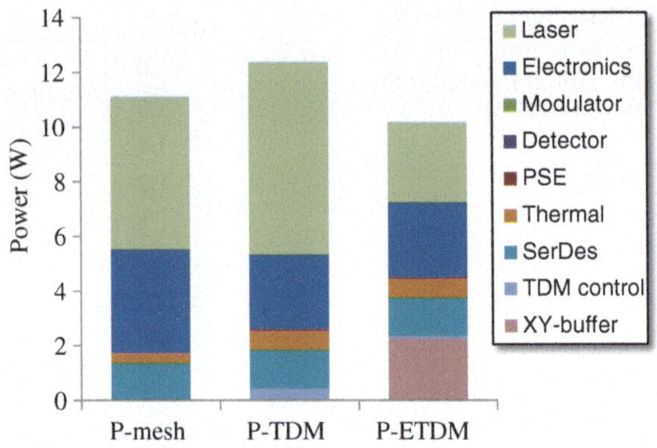

sustain the same bandwidth of the photonic links, enough electrical wires must be used to make up the difference between signaling rates, if any (in this case, 10 Gb/s photonic signaling, and 2.5 GHz for electronics). The TDM control circuitry contributes minimal power overhead to the two TDM networks. The P-ETDM network has lower insertion loss than P-mesh and P-TDM, and consequentially requires less laser power as well. However, the P-ETDM does dissipate additional power from the X-Y buffer (~2 W) which is used to store packets during the transition from X-dimension propagation to Y-destination propagation. Although P-ETDM has this additional source of power dissipation, it still consumes the lowest amount of overall total power.

7.2 Wavelength-Selective Spatial Routing

The TDM routing concept removes the costly overhead of circuit switching to provide a better performing interconnection network for small and unstructured traffic patterns (e.g. random). However, many applications executed within the scientific and cluster computing communities can still leverage circuit switching for better performance [12]. Typically, these applications require long-lived data connections to provide large amounts of data throughput. For this reason, circuit switching is still a relevant and important routing methodology.

This section describes wavelength selective spatial routing (WSSR) which is a method for enhancing the circuit switching technique by providing better network flexibility and routability.

7.2.1 Concept

Wavelength-selective spatial routing (WSSR) is used to selectively manipulate wavelength-channel subsets of a WDM signal as it propagates through a

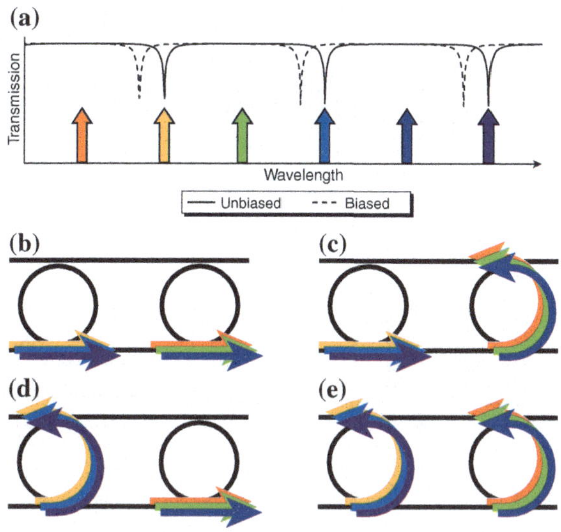

Fig. 7.12 **a** Spectral placement of two WDM partitions (each containing three wavelength channels), with respect to the spectrum of an electro-optic broadband ring switch. **b–e** Four possible routing configurations for a two-partition router [13]

network [13]. WSSR can be qualified as a hybrid form of spatial routing and wavelength routing (Fig. 5.2). The WSSR scheme takes advantage of the unused spectrum that exists between the resonances of a broadband ring switch by interleaving additional wavelength channels in the unused spectral space. The newly interleaved channels can then be used to provide additional paths of communication in the network to increase overall network performance.

Figure 7.12a illustrates the alignment of six wavelength channels relative to the spectral position of the modes of a single electro-optic ring switch. This differs from the traditional operation of a ring switch where wavelength channels are only positioned closely to the mode to enable switching. The example configuration includes a set of three wavelengths that are aligned to the resonances, but this WSSR-style configuration additionally includes a set of three wavelength channels tuned far away from any of the resonant modes. Each grouping of three wavelengths (mode aligned and unaligned) is called a *WDM partition*. These partitions are subsets of the total set of wavelengths available in the WDM system and form the basis for WSSR. WSSR has been experimentally demonstrated by Chan et al. [14].

The detuned WDM partition will propagate past the ring switch undisturbed, regardless of whether the ring is in its on-resonance or off-resonance state. However, a second ring resonator can be cascaded and designed to be aligned to the new set of wavelength channels forming a two-partition router. Introduction of the additional cascaded ring will increase the insertion loss by small amount. This small loss is attributed to the through-port loss of a ring switch. Figure 7.12b–e shows the four possible routing configurations of the 2-partition router, illustrating the independent controllability of each of the two WDM partitions.

Notice that in the previous example the original ring is augmented with a second ring resonator of the same diameter. This produces a wavelength channel spectrum

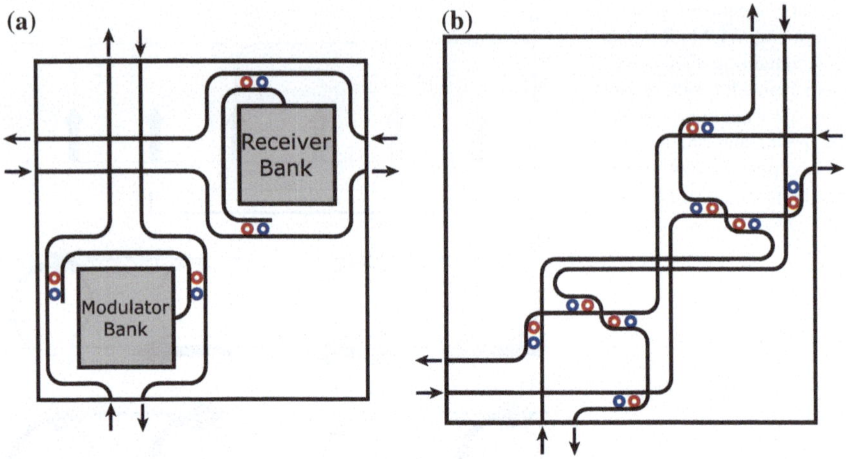

Fig. 7.13 Schematic of the TorusNX photonic routers configured with two WDM partitions: **a** gateway switch and **b** 4 × 4 non-blocking photonic switch [13]

that is effectively twice as dense as that of the original case. However, it may induce crosstalk due to the placement of wavelength channels close together. This design also has a more complex gateway since it requires a doubling of the number of modulator and photo-detector devices at each access point. Alternatively, the number of wavelength channels can be fixed to preserve the wavelength channel density and the rings can be designed to operate on a subset of the original wavelength channel set through an alteration of the FSR of the ring. A ring with half the diameter of the original will exhibit an FSR that is twice as wide and allows it to operate on half the original set of wavelength channels. This relationship between the number of partitions (and thus the number of rings) and the area footprint of the router is explored in Sect. 7.2.2.2.

The structure of the TorusNX switches (previously described in Sect. 5.3.2.3) configured with two WDM partitions is diagrammed in Fig. 7.13. Each pair of rings (indicated by a red and blue ring) composes the two cascaded rings that compose a two-partition router. Note that the original single-partition design of the gateway appears in Fig. 5.24 and can be reconstructed by removing either the red or blue set of rings from the layout. Similarly, the single-partition 4 × 4 non-blocking switch design in Fig. 5.8c can be reconstructed by removing one set of rings.

Preston, et al. found that a minimum wavelength channel spacing of 0.8 nm is required for ring modulated 10-Gb/s wavelength channels to maintaining sufficiently low crosstalk levels (< -20 dB) [15]. This corresponds to a 200-μm-diameter ring switch, which has been demonstrated previously with an adequately wide resonant modes for transmitting the high-speed datarate [16].

Smaller diameter microring switches can also be produced due to the fact that reductions in ring circumference will only reduce the circulating loss in the ring (though may increase radiation loss). It is possible that the reduced loss will increase the Q factor to a point where the drop-port resonance becomes too narrow to pass the high-speed data signal. This can be remedied by inducing additional insertion loss with fabricated defects or additional doping. The smallest demonstrated ring resonator device being consider in this analysis of WSSR has a diameter of 3 μm due to the dominance of bending losses [17]. The greatest number of WDM partitions considered in this analysis is six, which requires an 4.8-nm FSR and a 33.3-μm ring.

The ability to independently route each WDM partition enhances path diversity and forms the basis for WSSR. The number of WDM partitions is increased by interleaving additional sets of wavelengths, being only limited by the achievable wavelength channel density which must adhere to the aforementioned crosstalk constraints [15]. Single partition routers produce a degenerate case where the wavelength selectivity is eliminated, forming a purely spatially-routed design. Additionally, since the input-output port connectivity for all wavelength channels remains the same regardless of the number of WDM partitions, the entire router can be treated as a parameterized building block. These traits enable two features: (1) all spatially-routed topologies can be augmented with WSSR, and (2) the number of partitions and the network topology are independent design decisions that can be determined separately.

In a WSSR interconnect topology, each WDM partition can be regarded as an independent communication plane. This is conceptually analogous to electronic network multiplexing techniques such as traditional electronic virtual channels or the use of multiple physical networks. Implementation of traditional virtual channels is difficult in the photonic domain due to impracticality of optical buffering and processing. The use of multiple physical network planes is also detrimental since it will generate high insertion loss due to increased network complexity. Although increasing path multiplicity by adding extra paths in the network has been suggested [18], that analysis did not consider the fundamental physical-layer constraints of the network. These issues are circumvented with WSSR since the network planes are multiplexed in the wavelength domain.

Koohi, et al. have proposed 2D-HERT, a wavelength-routed network, which uses a similar partitioned wavelength space for directing wavelength channels [19]. The 2D-HERT network uses passive ring filters for guiding a subset of wavelengths. A source node employs source-routing through selection of an appropriate wavelength to establish the complete optical path since the wavelength determines whether the lightwave will pass through or drop into each passive ring filter. A characteristic feature of WSSR is the fact that it utilizes active electro-optic ring switches for generating several WDM partitions that act like independent network planes. The selection of wavelength only determines which network plane is traversed, but has no role in determining the optical path. An advantage of not utilizing wavelengths for routing purposes is that wavelength parallelism can be exploited for enabling higher node-to-node bandwidths. This type of network behavior is ideally suited for traffic with long-lived and large-message transmissions.

Allocation of WSSR network resources is accomplished using a circuit-switching methodology similar to the one used for spatial routing (Sect. 5.2). Processors interface with the network by communicating with a network gateway (Fig. 7.14).

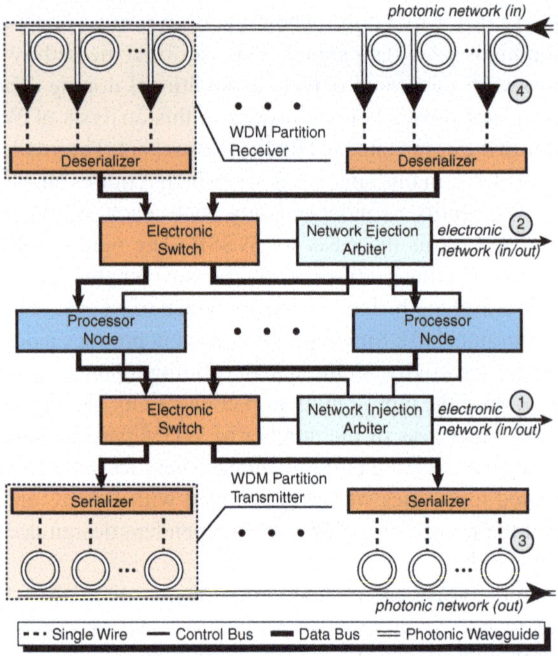

Fig. 7.14 The WSSR gateway architecture with concentrating processing cores [13]

Resource allocation of photonic routers is accomplished on a separate light-weight electronic packet-switched control plane, which has a topology that replicates the photonic layout.

The gateway performs the following tasks (enumerated in Fig. 7.14):

1. **Electronic/Transmission**: Processing cores first send transmission requests to the *Network Injection Arbiter* logic which handles the allocation of the *WDM Partition Transmitter* that implements the circuit-switching network protocol required to provision a photonic path.
2. **Electronic/Reception**: Requests from remote processing cores are sent to the *Network Ejection Arbiter* which handles the allocation of the *WDM Partition Receiver* that implements the circuit-switching network protocol for the reception end of the photonic link.
3. **Photonic/Transmission**: Each WDM Partition Transmitter is tuned to transmit on a different set of wavelengths, corresponding to a particular WDM partition.
4. **Photonic/Reception**: Each WDM Partition Receiver is tuned to receive on a different set of wavelengths, corresponding to a particular WDM partition.

The WSSR path-allocation protocol occurs through the transmission of a series of control messages on the electronic control plane. All control messages contain fields for message type, source ID, destination ID, and WDM partition selection data. The WDM partition selection field contains two flags (bits) per WDM partition that exists in the system. The 'check' indicates that a partition is being considered

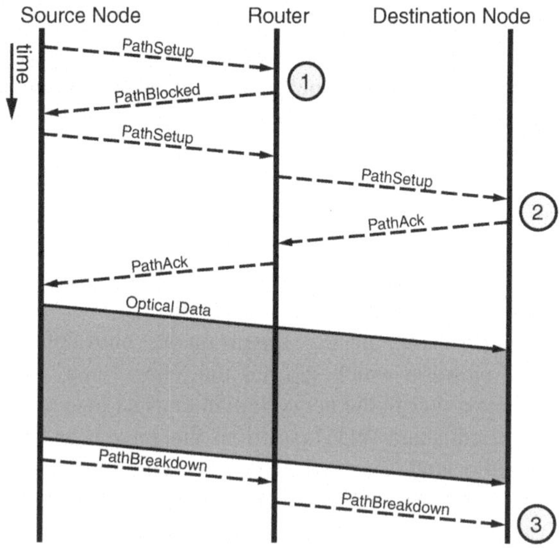

Fig. 7.15 Example timing diagram of the circuit-switching and WSSR allocation protocol. A path provisioning request is initially blocked, but is successful upon re-attempt [13]

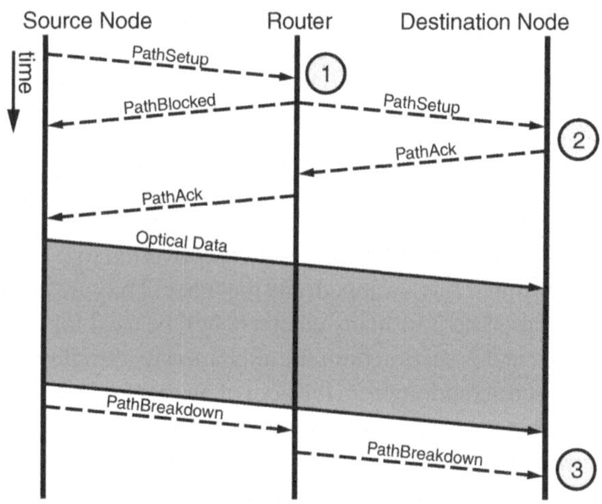

Fig. 7.16 Example timing diagram of the WSSR allocation protocol. If a single path provisioning request is attempted with multiple partitions, a path-setup request can partially block on a particular partition while be successful on another partition [13]

for allocation and the 'available' bit indicates that this resource is available for the corresponding partition. Figures 7.15 and 7.16 illustrate the message transactions required in perform allocation and data transmission. In the example, the request is initially blocked at an intermediate router, retries, and is subsequently successful in obtaining resource allocation and data transmission.

The allocation of a path begins with the transmission of a *PathSetup* from a source access point. The 'check' bit is set to 'true' on each partition for which allocation will be attempted. This automatically precludes partitions that have already been allocated from that particular source node and have not been de-allocated yet, or partitions where an allocation attempt is concurrently being made by another *PathSetup* message. Initially, the 'available' bits are all equal to the 'check' bits since an attempt at allocation is only performed if the partition is available at the gateway. For the simulation analysis presented in this section, the gateway only attempts to allocate a single WDM partition per *PathSetup* message. Alternatively, the *PathSetup* message could set a 'true' value for all 'check' bits which are free for allocation to increase the likelihood of finding a partition that is available. Other similar works have referred to the number of partitions used during each *PathSetup* as the *reservation aggressiveness* [20].

The *PathSetup* message travels on the control plane, attempting to provision each WDM partition which still has the 'check' and 'available' bit set as 'true'. Each photonic router in the network maintains its own reservation table, which is used to track circuits and WDM partitions that have been allocated or are in the process of being allocated. If any of the partitions are blocked, then a *PathBlocked* message is created and returned to the originating node, with the 'check' bit set to 'true' and 'available' bit set to 'false' for the blocked channels. The 'check' and 'available' bits that correspond to the blocked circuits are set to 'false' in the *PathSetup* message, and continue propagation only if at least a single WDM partition is still available. Figure 7.15 (marker '1') illustrates a situation where the path is blocked for all partitions being considered. Figure 7.16 (marker '1') shows a sequence of events where a subset of the available WDM partitions of the *PathSetup* message are blocked. The alternative WDM partition enables the *PathSetup* message to proceed and complete the provisioning process.

A *PathSetup* message that reaches the destination gateway indicates that at least one source-to-destination circuit is available for photonic transmission (Figs. 7.15 and 7.16 at marker '2'). The message is converted to a *PathAck* message, the source and destination ID are swapped, and the 'check' bits are preserved while the 'available' bit is set based on how many channels will be used for the transmission. The simulation studies in this section limit the allocation to a single WDM partition, and the partition is chosen at random from the pool of available channels as indicated by the 'available' bits. However, alternative configurations could enable some or all of the available partitions to be aggregated together to allow for dynamic throughput allocation.

Upon completion of the photonic transmission, a *PathTeardown* message is sent into the network from the source node (Figs. 7.15 and 7.16 at marker '3'). The 'check' bit is set for each partition that was allocated and is used to indicate to each photonic router along the path that the resources should be released and reservation table updated appropriately.

Table 7.2 Insertion loss parameters [13]

Parameter	Value	Ref.
Propagation loss (silicon)	1.7 dB/cm	[21]
Waveguide crossing	0.16 dB	[22]
Waveguide bend	0.005 dB/90°	[21]
Drop into a ring	0.6 dB	[16]
Pass by a ring	0.005 dB	[16]

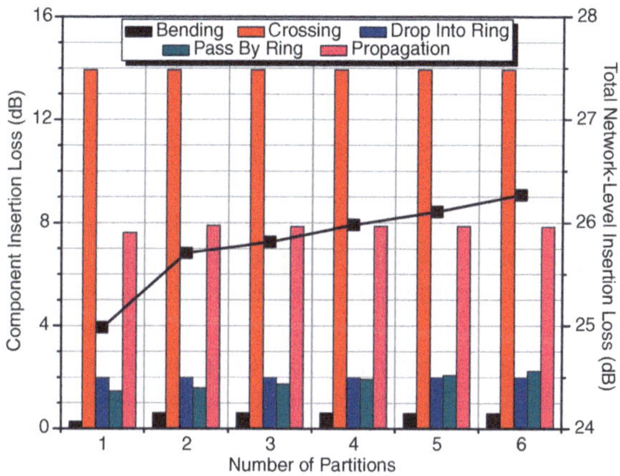

Fig. 7.17 Insertion loss analysis of the TorusNX topology for varying levels of partitioning. *Column plots* correspond to worst-case insertion loss per component among all possible network paths (*left-vertical axis*). The *line plot* corresponds to greatest total network-level insertion loss path among all possible network paths (*right-vertical axis*). The lossiest path does not necessarily correspond with the sum of the worst-case losses per component [13]

7.2.2 Analysis

7.2.2.1 Optical Power Budget and Insertion Loss Analysis

The consideration of the physical-layer properties of the photonic network plays a critical role in determining the feasibility of implementing the network. Specifically, the optical power budget and network-level insertion loss determine the requirements for the laser input power and for the receiver sensitivity. The insertion loss analysis assumes the parameters listed in Table 7.2, which are derived from experimentally-validated published results.

The results of the analysis are shown in Fig. 7.17 for different levels of partitioning. An initial cost of 0.72-dB insertion loss is observed when transitioning from one to two partitions; this jump in loss is attributed to additional waveguides and bends required to accommodate the additional ring resonators. Scaling beyond two partitions requires an increase in waveguide propagation and in the number of times ring resonators are passed, nonetheless a minor 0.56-dB loss increase is observed when transitioning from two to six partitions (0.14 dB per added partition).

The required laser power can be computed by adding the expected network loss to the receiver sensitivity. A receiver with a −17-dB sensitivity and operating at a 10-Gb/s datarate (demonstrated in [23]) would require a minimum injected laser power at the modulator of 8.0 dBm, 8.7 dBm, 8.8 dBm, and 9.0 dBm for one through four WDM partitions, respectively. Optical-power delivery mechanisms (Sect. 3.8) will need to be considered for inserting the required light into the NoC.

It is observed that the largest loss components arise from the waveguide crossings and the propagation. This shows that the introduction of WSSR into the photonic circuit-switching network topology only adds a small amount of loss to the network. The presented analysis assumes a planar single-crystalline silicon fabrication platform, but alternative CMOS-compatible platforms such as 3D deposited technology can virtually eliminate these loss constraints and increase the feasibility of this type of network [24].

7.2.2.2 Photonic Footprint

The nature of the WSSR mechanism requires multiple rings to enable the individual controllability of each WDM partition. As the number of WDM partitions increases, one of two ring design changes can be employed. In the first case, ring diameters are fixed regardless of the number of WDM partitions. This produces a system with higher channel density and consequentially higher wavelength channel crosstalk. Alternatively, the wavelength channel density can be fixed by scaling the ring diameter inversely proportional to the number of WDM partitions. While this has the benefit of not increasing spectral density of the channel spacing, it also enables a reduced footprint of the photonic routing element. The proceeding area analysis assumes the scaling of the ring diameters with a maximum considered diameter of 200 μm.

The area footprint of a single WSSR router versus the number of WDM partitions (labeled as the number of rings) is analyzed in Fig. 7.18. The WSSR router footprint calculations assume a structure similar to those shown in Fig. 7.12. The only locations where waveguides are closely placed together are regions where optical coupling is required (i.e. where the optical signal enters and exits the ring resonator). To prevent optical coupling across waveguides that are meant to be isolated, 5-μm gaps are used (e.g. between adjacent rings). An additional 2.5-μm gap is assumed to be on the outside edge of the two straight waveguides to account for space required with other optical components (e.g. another photonic router) outside of the immediate WSSR router of interest.

The plot shows the area scaling for varying initial single-partition ring diameters (as indicated in the legend). The curves show an immediate area benefit for increasing the number of partitions. Not only is this beneficial for the WSSR technique, but this scaling can also be used to benefit standard circuit-switching architectures through a reduction of the photonic footprint. The operational difference between WSSR and circuit switching is that the cascaded rings are used cohesively instead of independently. As the number of partitions increases, the area reduction diminishes and eventually an area increase is observed. The inflection point occurs at 40, 30, 20, and 10 rings for the 200, 150, 100, and 50 μm cases, respectively. Each curve

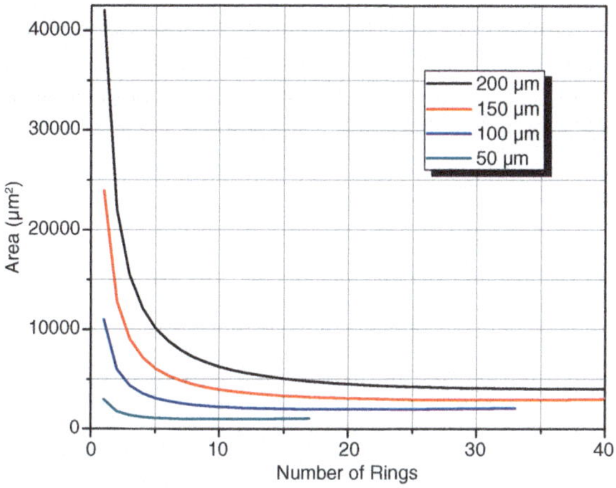

Fig. 7.18 Photonic router footprint for varying number rings (which corresponds to the number of WDM partitions enabled by the router). Legend indicates the ring diameter for the single ring case [13]

ends at the point where the individual ring diameters would become less than 3 μm which is the assumed minimum size limit of the ring resonators (corresponding to the smallest known fabricated microring [17]).

7.2.2.3 Contention Probability

From a performance perspective, the added path diversity by WSSR allows multiple communication links to occupy the same waveguides and photonic routers, resulting in reduced network-level contention. Decreased contentions will reduce latencies caused by network resource unavailability, and increase network-level bandwidth. Fundamentally, the use of multiple WDM partitions is equivalent to the concept of path multiplicity previously proposed and shown to improve performance of on-chip networks [18]. The primary difference in the two architectural concepts is in the usage of cascaded wavelength-selective spatially routed rings for WSSR and the overlay of additional waveguides and routing elements for path multiplicity. Studies have shown that waveguide crossings (which would be needed for added path multiplicity) are the largest contributor to insertion loss while the through port ring switch losses (used in WSSR) contribute a negligible amount [25]. This loss property is favorable for the WSSR design since it can be observed that the number of through port traversals will increase, but no additional crossing traversals will be created.

Destination contention occurs in the scenario when multiple source nodes request to transmit to a common destination node at the same time. This condition can occur within many traffic patterns where transmission requests experience hotspots. In the context of traditional circuit switching, a destination can only receive from a single source at any period in time. WSSR can alleviate this issue by providing multiple receiver connections for each destination.

A non-blocking network is assumed for this contention probability analysis. In a traditional circuit-switched non-blocking network, any idle source node can immediately transmit to its intended destination with the condition that the destination is not already receiving a message (i.e. no contention due to source blocking or circuit-path blocking). Consider a N node network, with a transmission being requested from source node A to destination node B. If all nodes aside from A have either established a connection or have been blocked (i.e. a saturated network), then there are $N-2$ nodes that could block this new connection. Assuming nodes do not require the optical network to communicate with itself, then the probability that the connection from A to B will not be destination blocked is

$$P_a = \left(\frac{N-2}{N-1}\right)^{N-2} \tag{7.5}$$

Next, a non-blocking WSSR network is considered with C WDM partitions. If each node is restricted to a single message transmission at a time (i.e. single transmitter per gateway), then the destination blocking probability of A is

$$P_C = 1 - \sum_{i=0}^{C-1} \binom{C}{i} P_a^{C-i} \cdot (1-P_a)^i \tag{7.6}$$

It can be easily shown that P_C will converge as $N \to \infty$, where the limits can be expressed exactly as

$$L_C = \lim_{N\to\infty} P_C = \left(1 - \frac{1}{e}\right)^C \tag{7.7}$$

Figure 7.19 plots Eq. 7.6 for $C = 1 \ldots 6$ and $3 \le N \le 100$, and appends the limit calculated from Eq. 7.7. A major improvement of the destination blocking probability is observed from $L_1 = 0.63$ to $L_6 = 0.064$. Furthermore, networks containing more than about 25 nodes vary minimally in terms of destination blocking probability. This indicates that techniques such as WSSR can improve the performance through a reduction in blocking probability.

7.2.3 Simulation Results and Analysis

The partitioned-WDM network architecture was modeled and simulated in the PhoenixSim design environment for a 2-cm × 2-cm CMP with 64 cores connect by an 8 × 8 network.

The photonic architectures have an electronic control plane operating with a 2.5-GHz clock. The control-plane routers utilize channel widths of 32 bit and 256-bit input buffers, corresponding to a buffer depth of 8 control messages. Path-setup control messages have an assumed bit length of 32 bit. Photonic networks are normalized by their total number of transmission wavelengths used, and wavelengths

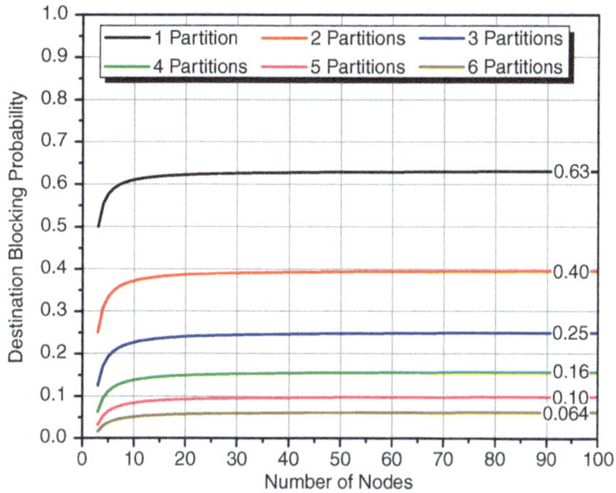

Fig. 7.19 Destination blocking probability in a non-blocking network for varying number of interleaved channels. The limit of each blocking probability as $N \to \infty$ is superimposed on the right of the plot [13]

are evenly allocated among the WDM partitions. Each wavelength channel provides a 10-Gb/s serial data rate. The TorusNX photonic circuit-switching topology design, augmented with wavelength-selective spatial routing, was used for this study [25].

A traditional electronic mesh network is also simulated to serve as a baseline comparison for the photonic architectures. Each electronic router has a channel width of 128 bit and utilizes a 2048-bit buffer on each input port. This electronic network model employs bubble flow control to prevent deadlocks. The electronic mesh network also operates on a 2.5-GHz clock, producing a link-level bandwidth of 320 Gb/s, and a network-level bisection bandwidth of 5.12 Tb/s for the 8×8 network.

Performance measurements were recorded for varying degrees of message size, total number of wavelength channels, and number of WDM partitions. Simulations were conducted with either a small (1-kbit) or large (100-kbit) message size. All synthetic traffic simulations utilized the standard uniform random traffic pattern. The number of WDM partitions ranged from 1 to 4 to capture the performance effect that the wavelength-selective spatial routing technique provides. The total number of wavelengths was varied between 12 (low aggressiveness), 60 (medium aggressiveness), and 120 wavelength channels (high aggressiveness).

Figure 7.20 contains plots for each combination of message size and total number of wavelength channels specified. The dotted-line curves depict the performance of the standard electronic mesh which is only influenced by the message size.

Photonic network configurations using small 1-kbit messages (left plots in Fig. 7.20) achieve saturation bandwidth gains that scale proportionally with the number of WDM partitions used. In the case of 60 and 120 wavelength channels, the small message sizes result in negligible differences in serialization delay when scaling the number WDM partitions. Consequently, this results in a fixed zero-load latency (approximately 90 ns) regardless of the number of WDM partitions, and saturation bandwidth gains that are approximately equal to the number of WDM partitions (e.g. 4 partitions results in a $4\times$ improvement). A difference in serialization delay

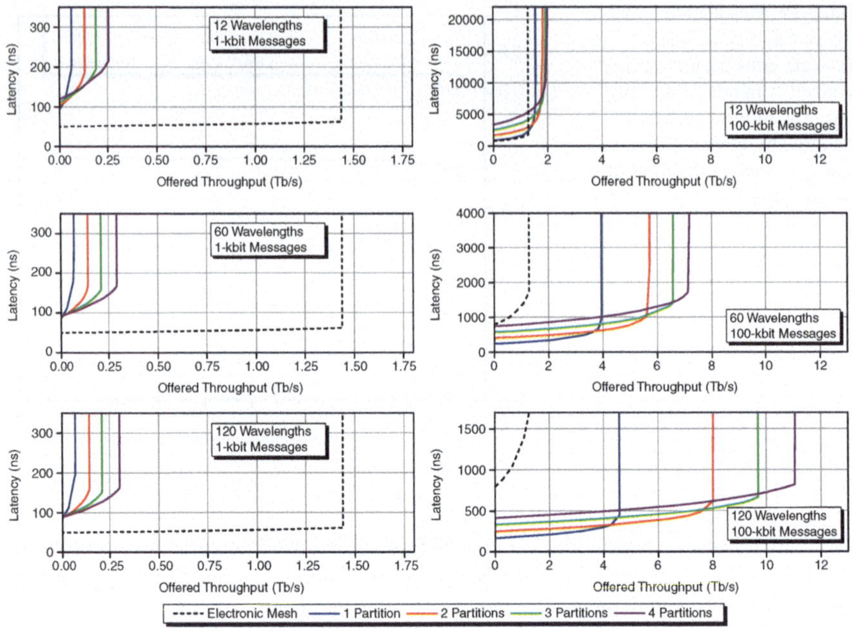

Fig. 7.20 Average latency versus offered throughput for varying number of WDM partitions, message sizes, and number of wavelength channels. Electronic mesh performance is shown as a *dotted line* [13]

is perceived only in the case of 12 wavelength channels. This results in a slightly degraded zero-load latency (120 ns for 4 partitions) and lower gain in saturation bandwidth (approximately 90 % gain per partition). The WDM partition technique provides significant performance gains relative to the degenerate case. The photonic network variants, however, still underperform in comparison to the electronic mesh, a disadvantage that has been previously observed in circuit-switched networks [7].

The transmission of 100-kbit messages (right plots in Fig. 7.20) on all the photonic network variants produce better performance values compared to the electronic mesh baseline. When compared to the degenerate case, the 12-wavelength system produces saturation-bandwidth gains of 14 %, 21 %, and 24 % when utilizing two, three, and four WDM partitions, respectively. In the 120-wavelength channel case, the saturation bandwidth gains are 97 %, 140 %, and 169 %, for the two, three, and four partition cases, respectively. In the best case, four partitions using a total of 120 wavelength channels achieve a saturation bandwidth improvement of 764 % over the electronic mesh. This shows that modest gains are achievable using WSSR for nearer term photonic networks, while greater gains can be expected as photonic device fabrication matures. Due to the large message sizes, the serialization delay is significantly longer and has a greater impact on the zero-load latency. For each set of plots with a common total wavelength count, the division of wavelength channels among WDM partitions produces noticeable differences in delay. In turn, this generates an interesting trade-off when determining whether a system design should minimize latency or maximize bandwidth.

References

1. G. Hendry et al., "Silicon nanophotonic network-on-chip using TDM arbitration," in *Proceedings of IEEE Symposium on High-Performance Interconnects*, Aug. 2010.
2. M. Millberg, E. Nilsson, R. Thid, A. Jantsch, "Guaranteed bandwidth using looped containers in temporally disjoint networks within the Nostrum network on chip," in *DATE '04: Proceedings of the conference on Design, automation and test in, Europe*, 2004, p. 20890.
3. K. Goossens, J. Dielissen, and A. Radulescu, "Æthereal network on chip: Concepts, architectures, and implementations," *IEEE Des. Test*, vol. 22, no. 5, pp. 414–421, 2005.
4. M. Schoeberl, "A time-triggered network-on-chip," in *International Conference on Field-Programmable Logic and its Applications (FPL 2007)*, Aug. 2007, pp. 377–382.
5. Z. Lu and A. Jantsch, "TDM virtual-circuit configuration for network-on-chip," *IEEE Trans. Very Large Scale Integr. Syst.*, vol. 16, no. 8, pp. 1021–1034, 2008.
6. C. Paukovits and H. Kopetz, "Concepts of switching in the time-triggered network-on-chip," in *RTCSA '08: Proceedings of the 2008 14th IEEE International Conference on Embedded and Real-Time Computing Systems and Applications*, 2008, pp. 120–129.
7. G. Hendry, J. Chan, S. Kamil, L. Oliker, J. Shalf, L. Carloni, and K. Bergman, "Silicon nanophotonic network-on-chip using TDM arbitration," in *2010 IEEE 18th Annual Symposium on High Performance Interconnects (HOTI)*, Aug. 2010, pp. 88–95.
8. A. Edman and C. Svensson, "Timing closure through a globally synchronous, timing partitioned design methodology," in *DAC '04: Proceedings of the 41st annual Design Automation Conference*, 2004, pp. 71–74.
9. J.-F. Zheng, F. Robertson, E. Mohammad, I. Young, D. Ahn, K. Wada, J. Michel, L. Kimerling, "On-chip optical clocking signal distribution," in *Optics in Computing*, 2003, p. OWB3.
10. G. Hendry, E. Robinson, V. Gleyzer, J. Chan, L. P. Carloni, N. Bliss, and K. Bergman., "Time-division-multiplexed arbitration in silicon nanophotonic networks-on-chip for high-performance chip multiprocessors." *Journal of Parallel and Distributed Computing*, Jan. 2011.
11. G. Roelkens, D. V. Thourhout, and R. Baets, "Continuous-wave lasing from DVS-BCB heterogeneously integrated laser diodes," in *Integrated Photonics and Nanophotonics Research and Applications*, 2007, p. ITuG4.
12. G. Hendry, S. Kamil, A. Biberman, J. Chan, B. G. Lee, M. Mohiyuddin, A. Jain, K. Bergman, L. P. Carloni, J. Kubiatowicz, L. Oliker, and J. Shalf, "Analysis of photonic networks for a chip multiprocessor using scientific applications," in *Proceedings of the 2009 3rd ACM/IEEE International Symposium on Networks-on-Chip*, ser. NOCS '09, May 2009, pp. 104–113.
13. J. Chan and K. Bergman, "Photonic interconnection network architectures using wavelength-selective spatial routing for chip-scale communications," *Optical Communications and Networking, IEEE/OSA Journal of*, vol. 4, no. 3, pp. 189–201, Mar. 2012.
14. J. Chan, N. Ophir, C. P. Lai, A. Biberman, H. L. R. Lira, M. Lipson, and K. Bergman, "Data transmission using wavelength-selective spatial routing for photonic interconnection networks," *Optical Fiber Communication Conference*, Mar. 2011.
15. K. Preston, N. Sherwood-Droz, J. Levy, and M. Lipson, "Performance guidelines for WDM interconnects based on silicon microring resonators," in *Lasers and Electro-Optics (CLEO), 2011 Conference on*, May 2011.
16. B. Lee, A. Biberman, P. Dong, M. Lipson, and K. Bergman, "All-optical comb switch for multiwavelength message routing in silicon photonic networks," *IEEE Photonics Technology Letters*, vol. 20, no. 10, pp. 767–769, May 2008.
17. Q. Xu, D. Fattal, and R. G. Beausoleil, "Silicon microring resonators with 1.5-μm radius," *Opt. Express*, vol. 16, no. 6, pp. 4309–4315, Mar. 2008.

18. A. Shacham, K. Bergman, and L. Carloni, "Photonic networks-on-chip for future generations of chip multiprocessors," *IEEE Transactions on Computers*, vol. 57, no. 9, pp. 1246–1260, Sep. 2008.
19. S. Koohi, M. Abdollahi, and S. Hessabi, "All-optical wavelength-routed NoC based on a novel hierarchical topology," in *Networks on Chip (NoCS), 2011 Fifth IEEE/ACM International Symposium on*, May 2011, pp. 97–104.
20. X. Yuan, R. Melhem, and R. Gupta, "Distributed path reservation algorithms for multiplexed all-optical interconnection networks," *Computers, IEEE Transactions on*, vol. 48, no. 12, pp. 1355–1363, Dec. 1999.
21. F. Xia, L. Sekaric, and Y. Vlasov, "Ultracompact optical buffers on a silicon chip," *Nature Photonics*, vol. 1, pp. 65–71, 2006.
22. W. Bogaerts, P. Dumon, D. V. Thourhout, and R. Baets, "Low-loss, low-cross-talk crossings for silicon-on-insulator nanophotonic waveguides," *OSA Optics Letters*, vol. 32, no. 19, pp. 2801–2803, 2007.
23. S. Assefa, B. G. Lee, C. Schow, W. M. Green, A. Rylyakov, R. A. John, and Y. A. Vlasov, "20 Gbps receiver based on germanium photodetector hybrid-integrated with 90 nm CMOS amplifier," in *in CLEO:2011 - Laser Applications to Photonic Applications*, 2011, p. PDPB11.
24. A. Biberman, K. Preston, G. Hendry, N. Sherwood-Droz, J. Chan, J. S. Levy, M. Lipson, and K. Bergman, "Photonic network-on-chip architectures using multilayer deposited silicon materials for high-performance chip multiprocessors," *J. Emerg. Technol. Comput. Syst.*, vol. 7, no. 2, pp. 7:1–7:25, Jul. 2011.
25. J. Chan, G. Hendry, A. Biberman, and K. Bergman, "Architectural exploration of chip-scale photonic interconnection network designs using physical-layer analysis," *IEEE/OSA Journal of Lightwave Technology*, vol. 28, no. 9, pp. 1305–1315, May 2010.

Chapter 8
Conclusions

Among the technologies that are emerging in the age of end-of-scaling CMOS, silicon photonics is perhaps the most promising to enable a smooth transition toward a new generation of post-CMOS computing systems. During the past decade a series of major breakthroughs in silicon photonic devices have demonstrated that all the components that are necessary to build chip-scale photonic interconnect components (e.g. modulators, filters, switches, detectors) can be fabricated using common CMOS processes. This key property of silicon photonics could allow a gradual integration of optical communication into CMOS integrated circuits. The photonics role can be increasingly expanded until it becomes central for the systems built with those so-called More-than-Moore technologies. Critically, during this transition silicon photonics could enable continued scaling of performance for a variety of applications because the unique properties of a chip-scale photonic network are not limited to the cross-chip communication distances. Instead, on-chip, board-scale, and cluster-scale distances are all equivalent in terms of optical communication performance. This is a fundamental difference with electronic communications, which must adhere to stricter bandwidth-distance product limitations.

On the other hand, the adoption of photonics into future computing systems is clearly not a one-to-one replacement. Its potentially transformational impact can be realized only after addressing some major technology challenges and by re-evaluating the composition of computing systems. Also, understanding how data movement in the optical domain can be leveraged to create new architectural paradigms will require the development of a new class of design automation methodologies and tools.

8.1 Major Technology Challenges Ahead

In this book we presented many silicon photonic devices and components that can be fabricated as discrete elements. However, the recent remarkable progress in the design and fabrication of each of these devices only scratches the surface of the scale

K. Bergman et al., *Photonic Network-on-Chip Design*,
Integrated Circuits and Systems, DOI: 10.1007/978-1-4419-9335-9_8,

of integration that is required for an effective utilization in computing systems. To demonstrate the viability of silicon photonics in full-scale systems, progress must be made towards higher integration density. For this to be successful, many silicon photonic devices must be shown to operate in harmony with each other, as well as with the complementary electrical components surrounding them.

Several fundamental challenges need to be addressed to achieve mass-scale operation. These include fabrication and performance uniformity, where silicon photonic devices would be capable of being produced with specified fabrication and performance tolerances in alignment with the rest of the semiconductor industry. For this to come to fruition, more emphasis must be placed on standardization and layout automation, especially when integrating silicon photonic devices with advanced microelectronics. This requires the creation of photonics-aware CAD tools that can be utilized by the semiconductor industry.

The inherent temperature dependence of silicon photonics produces another major scalability challenge, especially if silicon photonics is to be monolithically integrated within chip multiprocessors and systems-on-chip. An integrated circuit is intrinsically a dynamic thermal environment and the presence of an increasing number of heterogeneous components on the same chip augments the spectrum of temperature variations in both space and time. Hence, novel techniques for low-power thermal management of silicon photonic devices must be developed and standardized. This may include better methods of thermal compensation, or the development of compact and CMOS-compatible athermal devices.

Power dissipation is a challenge that researchers have started to address when demonstrating discrete silicon photonic devices. To have a major impact on the computing and communication industries, however, major performance gains must be achieved while simultaneously decreasing the consumption of power. This places tight constraints on the power dissipation of photonic devices. Denser integration of silicon photonics, especially with advanced microelectronics, will decrease the power consumed by both of these technologies. Intelligent power-management methods will ultimately need to be developed to enable the simultaneous operations of many photonic and electronic devices on the same integrated system.

Finally, a major challenge for silicon photonics is packaging. To truly benefit from the bandwidth density that photonics has to offer, novel packaging methods are necessary to connect on-chip photonic components to the outside world. For instance, the ability to connect two independent processors using an optical fiber is highly dependent on the robustness of the packaging solution. These methods must work beyond the carefully-controlled laboratory environment and be able to tolerate the uncertainties and harshness of the user environment.

None of these challenges represent a fundamental limitation but the collaborative effort of researchers and engineers working at the device, circuit, and system level is critical to make silicon photonic technology become a reality.

8.2 Integration and Scalability: The Role of Design Automation

Computer-aided design tools are necessary to realize the potential of photonic networks, enable their integration with electronic circuits, and sustain their scalability across system generations. The powerful and flexible VLSI CAD tools that exist for CMOS semiconductor technologies played a fundamental role in the proliferation of integrated circuits. In the same way, CAD tools for photonic systems will be necessary to leverage the capabilities that optical communication can provide. While some important progress has been made in recent years, a full comprehension of how to combine the basic photonic components into complete interconnection networks remains a challenging research area. This requires the development of a new class of models and tools for simulation, design-space exploration, performance analysis, physical design and optimization.

Tools for simulation and performance analysis create a pathway for modeling and evaluating many network architectures without having to go through a complete fabrication process. This is especially useful due to the fact that initial startup costs of chip fabrication can be very high. Also, since present fabrication technologies cannot be used to realize full-scale photonic chip designs, such tools allow engineers to investigate photonic architectures without the required technologies in place.

Tools for physical design and optimization make it possible to cope with the intrinsic complexity of assembling many devices in a correct way. These design automation tools have been instrumental for the continuous progress in the integration of electronic circuits, which nowadays host over a billion transistors. Similarly, they will be critical to derive the layout of systems that combine a mix of both electronic transistors and photonic devices, which have different properties and impose different constraints.

Finally, while different CAD tools need to be developed to solve specific problems at different levels of abstraction, it is critical to combine all these tools in a coherent design flow that can assist the engineering team throughout the design process from the system-level specification to the final physical implementation.

In conclusion, we believe that to realize the promise of chip-scale photonic communication requires a vertically-integrated research approach combining researchers in devices, networks, architectures, and applications. The design methodology and companion set of tools presented in this book represents a first step in this direction.

Index

K. Bergman et al., *Photonic Network-on-Chip Design*,
Integrated Circuits and Systems, DOI: 10.1007/978-1-4419-9335-9,

Q

R

S

T

The manufacturer's authorised representative in the EU is Springer Nature Customer Service Centre GmbH, Europaplatz 3, 69115 Heidelberg, Germany. If you have any concerns regarding our products, please contact ProductSafety@springernature.com

Printed and bound by CPI Group (UK) Ltd, Croydon, CR0 4YY
31/07/2026
02182826-0001